DAVID CLINTON

Keeping Up

Backgrounders to all the big technology trends you can't afford to ignore

Contents

Preface

"All the jargon. All the hype. All the crazy, over-the-top marketing pitches. And now: all in one place!"

Keeping Up is a quick, fun, and accessible guide to the current state of the tech industry and the big trends that will likely define its future. We'll discuss, at a high level, the major drivers of technology innovation and investment, and how it all interfaces with everything else. If you're responsible for making technology-based business decisions, looking for inspiration for new opportunities, thinking about a new career, or just curious about the world around you, getting all this information in one place can make for more intelligent and faster decisions.

Who might want to read this book?
- Managers and business decision makers
- Individuals plotting their career/educational trajectories
- Individuals looking for ideas/inspiration for new or undefined projects
- Curious people seeking to understand their world
- Journalists looking for quick context

Two images included in the book (oddly, both featuring versions of the Raspberry Pi computer) are made available through the Wikimedia Commons license. They are:

Figure 4.1: (The Raspberry Pi 4 Model B single-board mini computer) https://commons.wikimedia.org/wiki/File:Raspberry_Pi_4_Model_B_-_Side.jpg

And Figure 7.1: (The Raspberry Pi Zero - a fully-functioning server for under $10)

https://en.wikipedia.org/wiki/File:Raspberry-Pi-Zero-FL.jpg

Book cover by 100covers.com.

About the Author

David Clinton is an Amazon Web Services solutions architect and Linux server professional. He is the author of many books, including:

- AWS Certified Solutions Architect Study Guide: Associate SAA-C01 Exam 3rd Edition (Wiley/Sybex - with Ben Piper)
- AWS Certified Cloud Practitioner Study Guide: CLF-C01 Exam (Wiley/Sybex - with Ben Piper)
- Ubuntu Bible (Wiley - with Chris Negus)
- Linux Security Fundamentals (Wiley)
- Linux in Action (Manning)
- Learn Amazon Web Services in a Month of Lunches (Manning)
- Manage AWS Resources Using Ansible: the super short guide to cloud automation (Bootstrap IT)
- Solving for Technology: how to quickly learn valuable new skills in a madly changing technology world (Bootstrap IT)
- Teach Yourself Linux Virtualization and High Availability: prepare for the LPIC-3 304 certification exam (Bootstrap IT)
- Practical LPIC-1 Linux Certification Study Guide (Apress)

He also authored 25 video courses covering Linux and AWS administration, IT security, and security virtualization for Pluralsight.

You can reach David through his website, https://bootstrap-it.com.

I

The Big Picture

The next half dozen chapters will cover some of the biggest individual technology topics in some depth. You'll learn the basics behind security, data privacy, cloud computing, connectivity, and tech research.

1

Understanding Digital Security

Whatever your connection to technology, security should play a prominent role in the way you think and act. Technology, after all, amplifies the impact of everything we do with it:

- The things we say and write using communication technologies can be read and heard by many, many more people than would be possible without.
- The ability to conveniently connect with people and collaborate on projects of all kinds is much greater.
- The tasks we can perform are, through the magic of automation, almost limitless.
- The scope of information we can instantly access through the simplest and least expensive devices towers far beyond anything the greatest scholars could have hoped to see in a lifetime just a few decades ago.

All that means that criminals and other individuals unconstrained by moral conscience will have yet more powerful tools to compromise the data you create and consume, and steal or damage the property you acquire. So you've got a strong interest in learning how to protect yourself, your property, and that of the people and organizations around you.

This chapter will present a brief overview of what's at stake in the technology security domain. We'll define the kinds of threats we face and discuss the key tools at our disposable for pushing back against those threats.

Hacking? What's hacking?

Defining computer hacking in a way that doesn't anger someone, somewhere, is like talking about politics at work. Be prepared for long, awkward silences and possibly violence.

You see, purists might insist that the term hacking should apply exclusively to individuals focused on forcibly re-purposing computer hardware for non-standard purposes. Others reserve the title for people who bypass authentication controls to break into networks for criminal or political purposes. And how about those who wear the title as a sign of their practical expertise in all things IT? (And then, of course, there are crackers.)

But this is my book, so I'm going to use the term any way I want. I therefore decree that hacking is all about plans the *bad guys* have for *your* digital devices. Specifically, their plans to get in without authorization, get out without being noticed, and (sometimes) take your stuff with them when they leave. Using the term this way gives us a useful way to organize a discussion of some common and particularly scary threats.

How hackers get in

The trick is to find a way through your defenses (like passwords, firewalls, and physical barriers). In most cases, passwords probably provide the weakest protection:

- Passwords are often short, use a narrow range of characters, and are easy to guess.
- If a device came with a simple factory default password (like "admin" or "1234") just intended to get you in for the first time, then the odds are pretty good that many users will never get around to trading it in for something better.

- Even strong passwords can be stolen by deceptive phishing email scams ("Click here to login to your bank account..."); social engineering ("Hi, it's Ed from IT. We're having some trouble with your corporate account. Would you mind telling me your password over the phone so I can quickly fix it?"); and keyboard tracking software.

We'll talk more about firewalls later in this chapter. And physical barriers? I think you already know what a locked door looks like. But it's probably worth spending a few moments thinking about other kinds of digital attack.

The big prize is usually getting to your data and making off with copies. But for some, simply destroying the originals can be just as satisfying.

Obviously, logging into your devices using stolen passwords is the most straightforward approach. But access can also be achieved by intercepting your data as it travels across an insecure network.

One approach that's commonly used here is known as a man-in-the-middle attack, where data packets can be intercepted in transit and altered without authorized users at either end knowing anything's wrong. Properly encrypting your network connections (and avoiding unsafe public networks altogether) is an effective protection against this kind of threat. We'll talk more about encryption a bit later.

If the hardware you're using has an undocumented "back door" built in, then you're pretty much toast whatever you do. We'll talk more about back doors later in the book but, for now, I'll just note that there have been no shortage of factory-supplied laptops, rack servers, and even high-end networking equipment that's been intentionally designed to include serious access vulnerabilities. Be very careful where you purchase your compute devices.

If the attackers find a way into your physical building (sometimes posing as employees of a delivery company), they could quietly plug a tiny listening device into on unused ethernet jack on your network. That'll give them a nice platform to watch and even influence all your activities from the inside. Protecting your physical infrastructure and carefully monitoring network activity is your best hope against that kind of intrusion.

Even if your home or office is all fortressed up, there's no guarantee that data moving around on mobile devices (like smartphones or laptops) won't find its way into the wrong hands. And even if you've been careful to use only the best passwords for those devices, the data drives themselves can still be easily mounted as external partitions on a thief's own machine. Once mounted, your files and account information will now be wide open.

The only way to protect your mobile devices from this kind of threat is to encrypt the entire drive using a strong passphrase.

What hackers are after

Now that entire economies are run on computers directly connected to public networks, there's money and value to be had through well-planned corporate, academic, or political espionage efforts…and through old fashioned, traditional theft. Whether the goal is building up a military or commercial competitive advantage, completely destroying the competition, or just getting your hands on "free" money, illegally accessing other people's data has never been easier.

So what are hackers likely to be after? All the important financial and other sensitive information you'd prefer they didn't have. Including, it should be noted, the kind of information you use to identify yourself to banks, credit card companies, and government agencies.

Once the bad guys have got important data points like your birth date, home address, government-issued ID numbers, and some basic banking details, it's usually not hard for them to present themselves as though they're you, completely taking over your identity in the process.

Digital attacks can also be used as blackmail to force victims to pay to undo the damage that's been done. That's the objective of most *ransomware* attacks, where hackers encrypt all the data on a victim's computers and refuse to send the decryption keys needed to restore your rightful access unless you send them lots of money. Such attacks have already effectively brought down critical infrastructure like the IT systems powering hospitals and cities.

The very best defense against ransomware is to have full and tested backups of your critical data and a reliable system for quickly restoring it to your hardware. That way, if you're ever hit with a ransomware attack, you can simply wipe out your existing software and replace it with fresh copies, populated with your backed up data. But you should also beef up your general security settings to make it harder for ransomware hackers to get into your system in the first place.

When their primary goal is to prevent you or your organization from going about its business, hackers can remain at a safe distance and launch a distributed denial of service (DDoS) attack against your web infrastructure. Historical DDoS attacks have used massive swarms of thousands of illegally hijacked network-connected devices to transmit crippling numbers of requests against a single target service.

When large enough, DDoS attacks have managed to bring down even huge enterprise-scale companies using sophisticated defenses for hours at a time. The site hosting one of my favorite online open source collections was hit hard more than a year ago and still hasn't fully recovered.

What is encryption?

If your data is unreadable, there's a lot less bad stuff that unauthorized individuals will be able to do with it. But if it's unreadable, there's probably not a whole lot you'll be able to do with it either. Wouldn't it be nice if there was some way to present your data as unreadable in every scenario except where there's a legitimate reason? Well waddaya know? There is, and it's called data encryption.

Encrypting data in transit

Encryption algorithms encode information in a way that makes it hard, or even impossible, to be read. A simple (and ancient) example is symbol replacement, where every letter "a" in a message would be replaced with, say, the letter three positions on in the alphabet (which would be "d"). Every "b" would become "e" and so on. "Hello world" would be "khoor zruog". People subsequently coming across the message would be unable to understand it.

Of course, it wouldn't take long for a modern computer (or even a smart 8-year-old) to decode that one. But some very clever cryptologists have been working hard over most of the past century to produce much more effective algorithms. There are some significant variations of modern cryptography, but the general idea is that people can apply an encryption algorithm to their data and then safely transmit the encrypted copy over insecure networks so the recipient can then apply a decryption key of some sort to the data, restoring the original version.

Encryption is now widely available for many common activities, including sending and receiving emails. You can similarly ensure that the data you request from a website is the same data that's eventually displayed in your browser by checking the lock icon in your browser's address bar. The icon confirms that the website server employs Transport Layer Security (TLS) encryption.

Over the past few years, the Let's Encrypt project (letsencrypt.org) has encouraged millions of new websites to use encryption by providing free encryption certificates and simple-to-use tools to help server administrators install them.

Encrypting data at rest

TLS will protect your data when it's out and about, but what'll keep it safe even when it's relaxing in its comfy storage disk? File and drive encryption, that's what. All operating systems now offer integrated software for encrypting all or part of a storage disk either at installation time or later. Each time you power up an encrypted disk, you'll be prompted to enter the passphrase you created when you enabled encryption.

The thing is that if you forget your passphrase you're pretty much permanently locked out of your system and the data is as good as gone forever. But the other thing is that if you *don't* encrypt your system then, as we noted earlier, anyone who steals the hardware will have easy and instant access to your private information. It's a tough world out there, isn't it?

What does a firewall do?

You can think of a firewall as a filter. Just like, say, a water filter is able to block certain impurities, allowing only clean water through, a firewall can inspect every packet of data coming into or leaving your infrastructure, blocking access where appropriate.

Besides not needing to be replaced every few weeks, the big advantage of a firewall over a water filter is that it can be closely configured to permit and refuse entry to exactly match your security and functional needs, and then updated later should your needs change.

Hardware firewalls

A hardware firewall is a purpose-built physical networking device that's commonly used within enterprise environments. Such firewalls are installed at the edge of a private network and set to:

- Block potentially dangerous incoming traffic.
- Redirect other traffic to remote destinations.
- Permit traffic to access hosts within the local network.

Hardware firewalls are sold by companies like Cisco and Juniper, and general equipment manufacturers like HP and Dell. and can be used to manage traffic for networks encompassing many thousands of hosts. Firewalling appliances tend to be very expensive, often costing many thousands of dollars each. They're normally only deployed to manage enterprise infrastructure.

Software firewalls

A software firewall is an application that runs on a regular PC that can perform just about any function that you'd otherwise expect from a hardware firewall. There are two important differences:

- Firewall software (like the Linux iptables utility) is often free and, while complicated, enjoys the benefits of vast documentation resources. The software can also be installed on any old PC that's just lying around, reducing the overall cost to nearly nothing.
- You won't want to use such a firewall within a busy business environment however, since such a PC probably won't have the compute power to manage high volumes of network traffic. Nor, in most such cases, will it be reliable enough to provide mission-critical services 24/7.

There's another flavor of software firewall that's used as part of consumer-grade operating systems. Such firewalls allow you to better secure your OS by setting rules for what kind of activities you want to allow. These can be especially useful for mobile devices that frequently move from network to network.

Cloud computing platforms - like Amazon Web Services (AWS) and Microsoft's Azure - provide a firewall-like technology for use with the resources you might deploy within their systems. Firewall policies might exist in entities with names like "security group" or "access control list" that can be applied to whichever resource requires them.

Who does security best?

In the not too distant past, you would often hear IT professionals swearing they would never run their IT operations on infrastructure they didn't physically control. This was common when referring to outsourcing to third party, offsite companies or to cloud computing platforms. Whether it was because those administrators didn't trust the reliability and security of compute infrastructure run by strangers, or because regulatory restrictions required that sensitive workloads remained local, the sentiment was widely shared. And it made sense.

But the past is a different world. Today, it can be forcefully argued, the most secure and reliable environments are found in the biggest public cloud providers. Why? They've got the money and incentive to hire the very best engineers, and the money and incentive to build the very best infrastructure. Beyond that, cloud providers maintain data centers in political jurisdictions around the world, and go to great lengths to ensure their deployments comply with industry and government standards.

Let me illustrate. Remember a bit earlier in the chapter when we discussed DDoS threats? Well, back in the summer of 2020, an unnamed organization deploying resources on AWS was hit with a DDoS attack peaking at 2.3 Tbps. That is, each and every second, requests hit that organization's public-facing service with 2.3 terabytes of data.

What does "2.3 terabytes" actually mean? Well, a megabyte is (approximately) one million bytes of information (a PDF version of this book would probably take up six megabytes or so). A gigabyte is one thousand million bytes of information. A terabyte is one thousand thousand million bytes of information. That would be the equivalent of around 165,000 PDF books. 2.3 terabytes would be the rough equivalent of 380,000 PDF books.

Now try to imagine all the text characters used to fill 380,000 PDF books being thrown at a web service *each second.*

Got that image in your mind? So here's what happened to that web service: Nothing. It just carried on working as though it hadn't a care in the world. How on earth is that even possible? Amazon's AWS Shield service simply mitigated the attack. The customer didn't have to do a thing.

That is why moving your workloads to the public cloud doesn't necessarily involve compromising your standards.

Interested in digging deeper into this topic? My Linux Security Fundamentals book (Sybex, 2020) is entirely devoted to giving you the full picture. Even if you don't actually happen to work with Linux, there's enough platform-neutral content there to keep you good and busy.

2

Understanding Digital Privacy

Public service warning: you might find this chapter a wee bit depressing. If you'd prefer some cheering up right now, perhaps skip to "Understanding the Cloud."

For all the many benefits we enjoy from technology - and particularly the technologies that make up the public internet - there are clearly plenty of costs, too. Figuring out how you want to balance the benefits against the costs can take some careful thinking.

Here's a concise and effective way to describe the equation (whose source I've sadly forgotten):

> "Select any two of privacy, security, and convenience. But you can't have all three."

In other words, if security is a critical value for you, then you'll need to give up on 24/7 instant access to your money, credit, and personal accounts. That's because that kind of access requires exposing your accounts across public networks at a level that won't permit as much data protection as you might want.

Similarly, what if you just can't live without the convenience of getting news updates and social connectivity through sites belonging to third party businesses that collect and use your personal information? Well, you'll need

to "pay for it" by giving up a measure of your privacy.

Of course, most of us will choose some blend of those three elements based on a practical compromise between competing values and needs. But making a reasonable decision on that blend will require solid information. That's what you'll find through the rest of this chapter.

How companies get your data

Curious about what kinds of personal and even private data you may be exposing through the course of a normal day on the internet? How about using "all kinds" as a starting point? Perhaps the best way to understand the scope and nature of the problem is to break it down by platform.

Financial transactions

Take a moment to visualize what's involved in a simple online credit card purchase. You probably signed into the merchant's website using your email address as an account identifier and a (hopefully) unique password. After browsing a few pages, you'll add one or more items to the site's virtual shopping cart. When you've got everything you need, you'll begin the checkout process, entering shipping information, including a street address and your phone number. You might also enter the account number of the loyalty card the merchant sent you and a coupon code you received in an email marketing message.

Of course, the key step involves entering your payment information which, for a credit card, will probably include the card owner's name and address, and the card's number, expiry date, and a security code.

Assuming the merchant infrastructure is compliant with Payment Card Industry Data Security Standard (PCI-DSS) protocols for handling financial information, then it's relatively unlikely that this information will be stolen and sold by criminals. But either way, it will still exist within the merchant's own database.

To flesh all this out a bit, understand that using your loyalty card account and coupon code can communicate a lot of information about your shopping and lifestyle preferences. Not to mention records of some of your previous activities. Your site account comes with contact information and your home location.

All of that information can, at least in theory, be stitched together to create a robust profile of you as a consumer and citizen.

It's for these reasons that I personally prefer using third-party e-commerce payment systems like PayPal because such transactions leave no record of my specific payment method and on the merchant's own databases.

Devices

Modern operating systems are built from the ground up to connect to the internet in multiple ways. They'll often automatically query online software repositories for patches and updates and "ask" for remote help when something goes wrong.

Some performance diagnostics data is sent and stored online, where it can contribute to statistical analysis or bug diagnosis and fixes. Individual software packages might connect to remote servers independently of the OS to get their own things done.

All that's fine. Except that you might have a hard time being sure whether *all* the data coming and going between your device and the internet is stuff you're OK sharing. Can you know that private files and personal information aren't being swept in with all the other data? And are you confident that none of your data will ever accidentally find its way into some unexpected application lying beyond your control?

To illustrate the problem, I'd refer you to devices powered by digital assistants like Amazon's Alexa and the Google Assistant ("Ok Google"). Since, by definition, the microphones used by digital assistants are constantly listening for their key word ("Alexa…"), everything anyone says within range of the device is registered.

Figure 2.1: An Amazon Echo Show device that comes with an always-on microphone

At least some of those conversations are also recorded and stored online and, as it turns out, some of *those* have eventually been heard by human beings working for the vendor. In at least one case, an inadvertently-recorded conversation was used to convict a murder suspect (not that we're opposed to convicting murderers).

Amazon, Google, and other players in this space are aware of the issue and are trying to address it. But it's unlikely they'll ever fully solve it. Remember, convenience, security, and privacy don't work well together.

Now if you think the information from computers and tablets that can be tracked and recorded is creepy, wait 'till you hear about thermostats and light bulbs. As more and more household appliances and tools are adopted as part of "smart home" systems, more and more streams of performance data will be generated alongside them. And, as has already been demonstrated in multiple real-world applications, all that data can be programmatically interpreted to reveal significant information about what's going on in a home and who's doing it.

Mobile devices

Have you ever stopped in the middle of a journey, pulled out your smartphone, and checked a digital map for directions? Of course you have. Well the map application is using your current location information and sending you valuable information but, at the same time, you're sending some equally valuable information back. What kind of information might that be?

I once read about a mischievous fellow in Germany who borrowed a few dozen smartphones, loaded them up on a kids' wagon, and slowly pulled the wagon down the middle of an empty city street. It wasn't long before Google Maps was reporting a serious traffic jam where there wasn't one.

How does the Google Maps app know more about your local traffic conditions than you do? One important class of data that feeds their system is obtained through constant monitoring of the location, velocity, and direction of movement of every active Android phone they can reach - including your Android phone. I, for one, appreciate this service and I don't much mind the way my data is used. But I'm also aware that, one day, that data might be used in ways that sharply conflict with my interests. Call it a calculated risk.

Of course, it's not just GPS-based movement information that Google and Apple - the creators of the two most popular mobile operating systems - are getting. They, along with a few other industry players, are also handling the records of all of our search engine activity and the data returned by exercise and health monitoring applications.

In other words, should they decide to, many tech companies could effortlessly compile profiles describing our precise movements, plans, and health status. And from there, it's not a huge leap to imagine the owners of such data predicting what we're likely to do in the coming weeks and months.

Web browsers

Most of us use web browsers for our daily interactions with the internet. And, all things considered, web browsers are pretty miraculous creations, often acting as an impossibly powerful concierge, bringing us all the riches of humanity without even breaking into a sweat. But, as I'm sure you can already anticipate, all that power comes with a trade-off.

For just a taste of the information your browser freely shares about you, take a look at the Google Analytics page shown in figure 2.2.

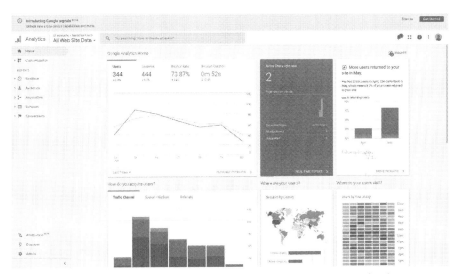

Figure 2.2: The home dashboard of a Google Analytics page displaying visualizations of visitors to a website

his dashboard displays a visual summary describing all the visits to my own bootstrap-it.com site over the previous seven days. I can see:

- Where in the world my visitors are from

- When during the day they tend to visit
- How long they spent on my site
- Which pages they visit
- Which site they left before coming to my site
- How many visitors make repeat visits
- What operating systems they're running
- What device form factor they're using (i.e., desktop, smartphone, or tablet)
- The demographic cohorts they belong to (genders, age groups, income groups)

Besides all that, a web server's own logs can report detailed information, in particular the specific IP address and precise time associated with each visitor. What this means is that, whenever your browser connects to my website (or any other website), it's giving my web server an awful lot of information. Google just collects it and presents it to me in a fancy, easy-to-digest format.

By the way, I'm fully aware that, by having Google collect all this information about my website's users that I'm part of the problem. And, for the record, I do feel a bit guilty about it.

In addition, web servers are able to "watch" what you're doing in real time and "remember" what you did on your last visit.

To explain, have you ever noticed how on some sites, right before you click to leave the page a "Wait! Before you go!" message pops up? Servers can track your mouse movements and, when they get "too close" to closing the tab or moving to a different tab, they'll display that popup.

Similarly, many sites save small packets of data on your computer called "cookies." Such a cookie could contain session information that might include the previous contents of a shopping cart or even your authentication status. The goal is to provide a convenient and consistent experience across multiple visits. But such tools can be misused.

Finally, like operating systems, browsers will also silently communicate with the vendor that provides them. Getting usage feedback can help providers stay up to date on security and performance problems. But independent tests have shown that, in many cases, far more data is heading back "home" than would seem appropriate.

Website interaction

Although some of this might be covered by previous sections in the chapter, I should highlight at least a couple of particularly relevant issues. Like, for instance, the fact that websites love getting you to sign up for extra value services. The newsletters and product updates that they'll send you might we be perfectly legitimate and, indeed, provide great value, but they're still coming in exchange for some of your private contact information. As long as you're aware of that, I've done my job.

A perfect example is the data you contribute to social media platforms like Twitter, Facebook, and LinkedIn. You may think you're just communicating with your connections and followers, but it actually goes much further than that.

Take a marvellous - and scary - piece of software called Recon-ng that's used by network security professionals to test for an organization's digital vulnerability. Once you've configured it with some basics about your organization, Recon-ng will head out to the internet and search for any publicly available information that could be used to penetrate your defenses or cause you harm.

For instance, Are you sure outsiders can't possibly know enough about the software environment your developers work with to do you any damage? Well perhaps you should take a look at the "desired qualifications" section from some of those job ads you posted on LinkedIn. Or how about questions (or answers) your developers might have posted to the Stack Overflow site? Every post tells a story, and there's no shortage of clever people out there who love reading stories.

Software like Recon-ng can help you identify potential threats. But that only underlines your responsibility to avoid leaving your data out there in public in the first place.

The bottom line? Smile. You're being watched.

Why companies want your data

Data is money. Some of the biggest and most successful tech companies of the past decade or two made their billions from data. Generally, that'd be from your data.

Of course, the value doesn't all move in one direction. Big tech companies do, as a rule, provide useful services. Health tracking apps do track and report on your health. Social media companies do (on rare occasion) provide for healthy and satisfying social interactions. And historical performance data does sometimes help improve customer and technical service.

But businesses exist to generate revenue and, as a rule, the more data they own, the more revenue it can generate. The more potential customers there are who provide their email and social media account coordinates, the easier it'll be to connect to them with new offers. And the easier it would be for other companies working in overlapping industries to connect to a business's customers as well. The incentive to sell contact lists to interested third party is obvious.

Naturally, legal restrictions and user agreements can sometimes stop such sales of data sets. But not every use-case is necessarily covered by such laws, and not every company is necessarily bound by a strong desire to follow the law.

A delicious case in point would be Canada's Do Not Call list from all the way back in 2004. The law prevented telemarketers from contacting anyone who had added themselves to the national list. The law required all telemarketers to remove all entries found on the list from their own call lists.

The problem was that spammers happily downloaded the Do Not Call lists and, confident that they represented confirmed active accounts, called those specifically. The only law that was effective in this case was the *law of unintended consequences*.

Your data can also be useful for personalizing the results you get from search engine queries. Of course, you might sometimes enjoy seeing results relating to previous browsing behavior, but don't lose sight of the fact that your behavior is being used as part of a campaign to sell you stuff.

It's not only search engines: smartphone browsing histories are sometimes used by nearby businesses to push customized ads in your direction - sometimes even through automated digital displays on physical billboards and other signage.

Perhaps the biggest value your data can offer is when it's aggregated along with data generated by thousands or millions of other users. Data scientists can stream and parse huge, dynamic data sets to extract significant insights about subtle but significant trends. In many cases, such data is sanitized to remove any personally identifiable information (PII).

We can nicely sum up the 21st Century web application business model with this popular - and accurate - expression:

"If you're not paying for the product, you are the product."

How to protect your data

All that sounds pretty bleak. After all, George Orwell's 1984 was meant to be a warning, not a how-to guide. What can you do to push back?

Be aware of your environment.

Do you still even notice those terms of service disclosures you "click to sign" before they'll let you use some service or tool? Some of those disclosures are as long as this chapter - and, if I may say so myself, a whole lot less fun. But the fact is that they contain information that can have a profound impact on both you and your data.

Many agreements describe what data they're likely to collect and what they're planning to do with it. They'll often also offer assurances that they'll never sell your data to third parties - an assurance that they might sometimes even honor in both the letter and the spirit of the law (although there have been famous cases of companies that did neither).

I've never met anyone who has the time and energy to read through those endless disclosures from end to end. But if an organization pays a bunch of lawyers to write something, you can bet it's a serious business.

Be aware of your rights.

Beyond your specific agreement with a technology service provider, the use of your data might be regulated by government legislation. One example is the European Union's General Data Protection Regulation (GDPR), which controls how organizations must treat any personal data they encounter in the course of their operations.

Another example is the US government's Health Insurance Portability and Accountability Act (HIPAA), which regulates the handling of private information in the health insurance and healthcare industries.

Be aware of your alternatives.

Consider adopting privacy-first tools instead of the more heavily commercial services you're using now. For instance, the DuckDuckGo.com search engine, whose home page is shown in figure 2.3, doesn't track your search behavior and will return the same results to a particular query for you as for anyone else. They are a for-profit business, but they earn much of their income through affiliate links that pay them a commission for sales generated through search links - none of which has any impact on your privacy.

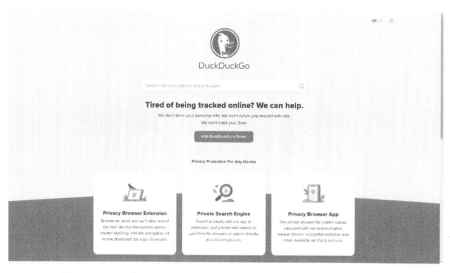

Figure 2.3: The home page of the DuckDuckGo search engine

The Brave browser, as another example, has been shown to send far less undocumented data out to the internet than any other major browser. To be specific, in early 2020, Douglas Leith of the School of Computer Science

& Statistics, Trinity College Dublin, tested six browsers for their risks of revealing unique identifying information about their host computers (scss.tcd.ie/Doug.Leith/pubs/browser_privacy.pdf). He found that Brave clearly offered the greatest privacy protection.

Brave also blocks web page ads by default, which raises a question. Since many web pages earn income exclusively through display ads, does Brave expect content providers to offer their services for free?

The browser provider actually has a business model that includes the content providers: users of the Brave browser can opt to be shown simple and extremely unobtrusive ads from carefully curated advertisers in exchange for micro payments in a crypto currency. The users can then choose to make micro payments to website content providers using those funds as a way to pay for their content through the Brave Rewards program (pictured in figure 2.4).

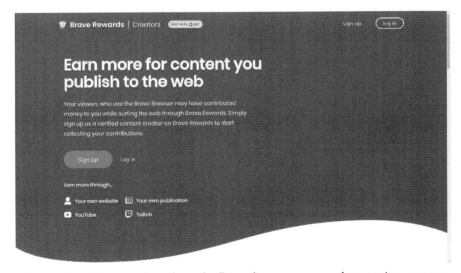

Figure 2.4: Web page describing the Brave browser approach to paying content providers

Opting for open source applications can also be an effective privacy strategy. OpenStreetMap (openstreetmap.org) is an alternative to Google Maps. It might not have all the bells and whistles and built-in connectivity you may be used to, but it's just that kind of connectivity that powers our concerns, isn't it?

If you're not comfortable with the big mobile operating system players (Android and iOS), you could, instead, buy a phone and install one of a number of experimental mobile Linux variations. But bear in mind that going down this road will likely be bumpy. Expect to run into unexpected configuration and compatibility challenges, and don't expect to find all the convenient apps that you've come to know and love using the big app stores.

See a hole that needs filling? Why not contribute your own innovation by participating in existing open source projects or adding your own solutions to the community?

3

Understanding the Cloud

You may not always be aware of it, but you're enjoying the many fruits of the cloud just about every hour of every day. Many of the joys (and horrors) of modern life would be impossible without it. Before we talk about what it does and where it's taking us, we should explain exactly what it is.

The "cloud" is all about using other people's computers rather than your own. That's it. No, really.

Cloud providers run lots of compute servers (which are just computers that exist to "serve" applications and data in response to external requests), storage devices, and networking hardware. Whenever the impulse takes you, you can provision units of those servers, devices, and networking capacity for your own workloads. When you add millions more users taken by similar impulses, you get the modern cloud.

For many - although not all - applications, there are enormous cost and performance benefits to be realized by deploying to a cloud. And countless applications - whether small, large, or smokin' colossal - have found productive homes on one cloud platform or another. So let's see how it all works and what you might be able to do with it.

Application server deployment models

Over the decades, we've been through a number of models for running server workloads. In a way, all those changes have been the product of just two technologies:

- Networking protocols that permit communication between connected nodes
- Virtualization which permits fast, efficient, and cost-effective use of hardware resources for multiple and parallel uses

Networking, largely because it's now such a stable and well established technology, isn't something we'll focus on here. But we will get back to virtualization a bit later.

Local data centers

In the old days, if you wanted to fire up a new server to perform a compute task, you would spend a week or so calculating how much compute power you'd need for your job, contact the sales reps at a few hardware vendors, wait for them to get back to you with bids, compare the bids and, when you've selected one, wait another couple of weeks for your new hardware to be delivered. Then you'd put all the pieces together, plug it all in, and start loading software.

The room where your servers ran would need a reliable and robust power supply and some kind of cooling system (Like angry children, servers generate a great deal of heat but don't like being hot.)

You probably wouldn't want to do any other work in that room, since the noise of your servers' powerful internal cooling fans was difficult to ignore.

While locally-deployed servers gave you all the direct, manual control over your hardware that you could need, it came at a cost. For one thing, opportunities for infrastructure redundancy (and the reliability that comes with it) were limited. After all, even if you regularly backed up your data (and assuming your backups were reliable), they still wouldn't protect you from a facility-wide incident like a catastrophic fire. You would also need to manage your own networking, something that could be particularly tricky - and risky - when remote clients required access from beyond your building.

By the way, don't be fooled by my misleading use of past tense here ("were limited," "backed up"). There are still plenty of workloads of all sizes happily spinning away in on-premises data centers. But the trend is, without question, headed in the other direction.

Server co-location

Another option for mid-sized to large organizations is to store your own servers in someone else's data center, an arrangement known as co-location. The hosting company provides the server racks and power, along with all the networking and cooling equipment you'll need. Whenever you need physical access to your servers, they'll always be happy when you drop by to say hello.

This is a convenient way to maintain direct control over your servers while leaving physical security and the larger infrastructure headaches in the hands of specialists. Co-location facilities are often capable of far higher standards of security and reliability then smaller operations would be able to manage on their own.

For security reasons, co-location centers will probably not advertise their services at street level. But if you want to see what they look like, search for "server hosting co-location" in your city and then use Google Satellite to check out one or two of the addresses that come back. If you see a large, unmarked building with dozens of powerful air conditioning units on the roof, that'll be a data center.

Virtualization

As I hinted earlier, virtualization is the technology that, more than any other, defines the modern internet and the many services it enables. At its core, virtualization is a clever software trick that lets you convince an operating system that it's all alone on a bare metal computer when it is, in fact, just one of many OSs sharing a single set of physical resources. A virtual OS will be assigned space on a virtual storage disk, bandwidth through a virtual network interface, and memory from a virtual RAM module.

Here's why that's such a big deal. Suppose the storage disks on your server host have a total capacity of two terabytes and you've got 64GB of RAM. You might need 10GB of storage and 10GB of memory for the host OS (or, *hypervisor* is some virtualization hosts are called). That leaves you a lot of room for your virtual operating system instances.

You could easily fire up several virtual instances, each allocated enough resources to get their individual jobs done. When a particular instance is no longer needed, you can shut it down, releasing its resources so they'll instantly be available for other instances performing other tasks.

But the real benefits come from the way virtualization can be so efficient with your resources. One instance could, say, be given RAM and storage that, later, proves insufficient. You can easily allocate more of each from the pool - often without even shutting your instance down. Similarly, you can reduce the allocation for an instance as its needs drop.

This takes all the guesswork out of server planning. You only need to purchase (or rent) generic hardware resources and assign them in incremental units as necessary. There's no longer any need to peer into the distant future as you try to anticipate what you'll be doing in five years. Five *minutes* is more than enough planning.

Now imagine all this happening on a much larger scale. Suppose you've got many thousands of servers running in a warehouse somewhere that are hosting workloads for thousands of customers.

Perhaps one customer suddenly requests another terabyte of storage space. Even if the disk that customer is currently using is maxed out, you can easily add another terabyte from some other disk, perhaps one plugged in a few hundred meters away on the other side of the warehouse. The customer will never know the difference, but the change can be nearly instant.

Cattle vs pets

Server virtualization has changed the way we look at computing and even at software development. No longer is it so important to build configuration interfaces into your applications that'll allow you to tweak and fix things on the fly. It's often more effective for your developers and sysadmins to build a custom operating system *image* (nearly always Linux-based) with all the software pre-set. You can then launch new virtual instances based on your image whenever an update is needed.

If something goes wrong or you need to apply a change, you simply create a new image, shut down your instance, and then replace it with an instance running the new image. Effectively, you're treating your virtual servers the way a dairy farmer treats cows: when the time comes (as it inevitably will), you take an old or sick cow out and kill it, and then bring in another (younger) one to replace it.

Anyone who's ever been involved with legacy server room administration would gasp at such a thought! Our old physical machines would be treated like beloved pets. At the slightest sign of distress, we'd be standing, concerned, at its side, trying to diagnose what the problem was and how it can be fixed. If all else failed, we'd be forced to reboot the server, hoping against hope that it came back up again. If even *that* wasn't enough, we'd give in and replace the hardware.

But the modularity we get from virtualization gives us all kinds of new flexibility. Now that hardware considerations have been largely abstracted out of the way, our main focus is on software (whether entire operating systems or individual applications).

And software, thanks to scripting languages, can be automated. So using orchestration tools like Ansible, Terraform, and Puppet, you can automate the creation, provisioning, and full life cycle management of application service instances. Even error handling can be built into your orchestration, so your applications could be designed to magically fix their own problems.

Virtual machines vs containers

Virtual instances come in two flavors. Virtual machines (or VMs) are complete operating systems that run on top of - but to some degree independent of - the host machine. This is the kind of virtualization that uses a hypervisor to administrate the access each VM gets to the underlying hardware resources.

Such VMs are generally left to live whichever way they choose. Examples of hypervisor environments include the open source Xen project, VMware ESXi, Oracle's VirtualBox, and Microsoft Hypver-V.

Containers, on the other hand, will share not only hardware, but also their host operating system's software kernel. This makes container instances much faster and more lightweight (since their images don't need to include a kernel).

Not only does this mean that containers can launch nearly instantly, but that their file systems can be transported between hosts and shared. Portability means that instance environments can be reliably reproduced anywhere, making collaboration and automated deployment not only possible, but easy.

Examples of container technologies include LXD and Docker. And enterprise container implementations include Google's open source Kubernetes orchestration system.

Public clouds

Public cloud platforms have elevated the abstraction and dynamic allocation of compute resources into an art form. The big cloud providers leverage vast networks of hundreds of thousands of servers and unfathomable numbers of storage devices spread across data centers around the world.

Anyone, anywhere, can create a user account with a provider, request an instance using a custom-defined capacity, and have a fully-functioning and public-facing web server running within a couple of minutes. And since you only pay for what you use, your charges will closely reflect your real-world needs.

A web server I run on Amazon Web Services (AWS) to host two or three of my moderately busy websites costs me only $50 a year or so and has enough power left over to handle quite a bit more traffic. The AWS resources used by the video streaming company Netflix, will probably cost a bit more - undoubtedly in the millions of dollars per year. But they obviously think they're getting a good deal and prefer using AWS over hosting their infrastructure themselves.

Just who are all those public cloud providers, I'm sure you're asking? Well that conversation must begin (and, often, end) with AWS. They're the elephant in every room. The millions of workloads running within Amazon's enormous and ubiquitous data centers, along with their frantic pace of innovation, make them the player to beat in this race. And that's not even considering the billions of dollars in net profits they pocket each quarter.

At this point, the only serious competition to AWS are Microsoft's Azure which is doing a pretty good job keeping up with service categories and, by all accounts, is making good money in the process; and China's Alibaba Cloud which is mostly focused on the Asian market at this point. Google Cloud is in the game, but appears to be focusing on a narrower set of services where they can realistically compete.

As the barrier to entry in the market is formidable, there are only a few others who are getting noticed, including Oracle Cloud, IBM Cloud and, with a welcome nod to creative naming choices, Digital Ocean.

Private clouds

Cloud goodness can also be had closer to home, if that's what you're after. There's nothing stopping you from building your own cloud environments on infrastructure located within your own data center. In fact, there are plenty of mature software packages that'll handle the process for you. Prominent among those are the open source OpenStack (openstack.org) and VMware's vSphere (vmware.com/products/vsphere.html) environments.

Building and running a cloud is a very complicated process and not for the hobbyist or faint of heart. And I wouldn't try downloading and testing out OpenStack - even just to experiment - unless you've got a fast and powerful workstation to act as your cloud host server and at least a couple of machines for nodes.

You can also have it both ways by maintaining certain operations close to home while outsourcing other operations in the cloud. This is called a hybrid cloud deployment. Perhaps, as an example, regulatory restrictions require you to keep a backend database of sensitive customer health information within the four walls of your own operation, but you'd like your public-facing web servers to run in a public cloud. It's possible to connect resources from one domain (say, AWS) to another (your data center) to create just such an arrangement.

In fact, there are ways to closely integrate your local and cloud resources. The *VMware Cloud on AWS* service makes it (relatively) easy to use VMware infrastructure deployed locally to seamlessly manage AWS resources (aws.amazon.com/vmware).

The value of outsourcing your compute operations

Why might you want to migrate workloads to the cloud? You could end up saving a lot of money. So there's that. Of course, it's not going to work out that way for every deployment, but there do seem to be a lot of use cases where it does.

To help you make informed decisions, cloud platforms often provide sophisticated calculators for you to compare the costs of running an application locally as opposed to what it would cost in the cloud. The AWS version of that is here: aws.amazon.com/tco-calculator

Part of the pricing calculus is the *way* you pay. The traditional on-premises model involved large up-front investments for expensive server hardware. The hope was that your investment would deliver enough value over the next five to ten years to justify the purchase. Those purchases are known as *capital expenses* ("Capex").

Cloud services, on the other hand, are billed incrementally (by the hour, or even minute) according to the number of service units you actually consume. This is normally classified as *operating expenses* (Opex).

Using the Opex model, if you need to run a server workload only once every few days for five minutes at a time in response to an external triggering event, you can automate the use of a "serverless" workload (using a service like Amazon's Lambda) to run only when needed. Total costs: perhaps only a few pennies a month to cover those minutes the service is actually running.

Besides cost considerations, there's a lot more going on in the cloud world that should attract your consideration. You've already seen how the lag time between the decision to deploy a new server on-premises and its actual deployment (weeks or months) compares to a similar decision-to-deployment process in a public cloud (a few minutes). But large cloud providers are also positioned to deliver environments that are significantly more secure and reliable.

As an example, you may remember our story about the DDoS attack from chapter 2 (Understanding Digital Security). That was the incident where the equivalent of 380,000 PDF books worth of data were used to bombard an AWS-hosted web service each second...and the service survived. Are you confident you could do that yourself?

And how about reliability through redundancy? Would your on-premises infrastructure survive a catastrophic loss of your premises? Even if you did the right thing and maintained off-site backups, how long would it take you to apply them to rebuilt, network-connected, and functioning hardware?

The big cloud platforms run data centers across physically distant locations around the world. They make it easy (and in some cases unavoidable) to replicate your data and applications in multiple locations so that, even if one data center goes down, the others will be fine. Can you reproduce that?

Cloud providers also manage content distribution networks (CDNs) allowing you to expose cached copies of frequently-accessed data at edge locations near to wherever on earth your clients live. This greatly reduces latency, improving the user experience your customers will get. Is *that* something you can do on your own?

One more thought. Most of the big investments into new IT technologies these days are being plowed into cloud ecosystems. That's partly because the big cloud providers are generating cash far faster than they can hope to spend it, but it's also because they're involved in a live-or-die race to capture new segments of the infrastructure market before the competition claims them.

The result is that the sheer rate of innovation in the cloud is staggering. I earn a living keeping a close eye on AWS, and even I regularly miss new product announcements. One of the reasons I avoid including screenshots of the AWS management console in my books and video courses is because their console is updated so often, the images will often be out-of-date before the book hits the street.

In some cases, this might mean that local deployments will run at a built-in disadvantage simply because they won't have access to the equivalent cutting edge technologies.

The risks of outsourcing your compute operations

Having said all that, as with most things in life, choosing between cloud and local isn't always going to be as obvious as I may have made it sound. There may still be, for instance, laws and rules forcing you to keep your data local. There will also be cases where the math just doesn't work out: sometimes it really is cheaper to do things in your own data center.

You should also worry about platform lock-in. The learning curve necessary before you'll be ready to launch complex, multi-tier cloud deployments isn't trivial. And you can be sure that the way it works on AWS, probably won't be quite the same as what's happening on MS Azure. The knowledge investment you'll need to make once you make your choice will probably be expensive.

And then what happens to that investment if the provider's policies suddenly change in a way that forces you off the platform? Or if they actually go out of business? Could happen: Kodak, Blockbuster Video, and Palm were once big, too.

And what about getting locked out of your account for some reason? How hard would it be for you to retool and reload everything somewhere else?

Just think ahead and make sure you're making a rational choice.

4

Understanding Digital Connectivity

Telephones changed the way we all talked to each other and went about our work (well, the way our great-grandparents did, at any rate). Information could now be communicated instantly, rather than being sent over slow, overland routes. But that's hardly news to anyone these days. The modern network - best known through its *internet* iteration - similarly boosted communication, although this time it was the movement of *data* rather than *voice* that got a boost.

In the fifty years or so since the birth of the internet, it's been trusted with the movement, storage, and management of more and more of our data. These changes have brought tremendous opportunities, risks, and pressures. Just getting connected is now a basic necessity. Managing all of our many connected devices and leveraging the ways we authenticate to extend our identities also present challenges. We'll discuss all that in this chapter.

Connecting to the internet

These days, after food and shelter, pretty much the most basic resource of all is internet connectivity. If you can't access the internet, you'll find it difficult to do your banking, educate yourself, book travel arrangements, or even figure out exactly where you are. It's not for nothing that widespread, reliable, and relatively fast internet access is critical for a region's general economic development.

Even though the internet was originally build as a decentralized, distributed network of resources, you still need to establish some kind of connection to access it. The best connections are run by network carriers, known as *tier 1 networks*. These networks can reach all other networks through a peering arrangement that doesn't require payment for IP transit. You can think of these networks as the backbone of the internet, and their network infrastructure as the rules that define its shape.

Examples of companies managing tier 1 networks include AT&T and Verizon in the US, Tata Communications (India), and Deutsche Telekom (Germany). Those carriers will resell bandwidth to smaller internet service providers (ISPs) who, in turn, sell access to end users like you and me.

Broadband options

Individuals looking for broadband access in their homes or small businesses can usually choose between one of four access models:

- **Cable**. Since they're already in the business of providing data to millions of homes over existing physical connections, cable TV providers can easily transmit internet over the same wires.
- **Digital subscriber line (DSL)**. A family of technologies that permit the movement of digital data across copper telephone lines, DSL can provide a roughly similar level of service as cable, but without the need for an underlying cable subscription. In fact, using a "dry copper" connection, you don't even need a telephone landline account.
- **Fibre optics**. Due to some arcane technical details (including the laws of physics), transmitting digital signals as infrared light can happen faster and require fewer repeaters than comparable electrical cables. A fibre optics internet connection could typically deliver transfer speeds of 10-40Gbit/s - a thousand times faster than currently standard rates using cable or DSL. Although your ISP might choose to cap your actual connection at much lower speeds. Always read the fine print on the

contract.

- **Satellite.** Running new cable through densely populated cities is expensive, but companies can quickly make their money back through the many access contracts they'll sign. But sparsely populated rural regions are much more difficult to service. Partly to fill the rural connectivity gap, a number of companies are ambitiously working to launch constellations of thousands of orbiting satellites to provide universal internet coverage. As of this writing, SpaceX is furthest along with its project, having already successfully launched more than 500 satellites as part of the Starlink system.

Besides those dominant technologies, there have been more than a few alternate connectivity solutions attempted. Some are experimental but promising, and others are a bit more speculative. Google's Balloon Internet (known officially as Loon LLC), is an attempt to float fleets of high-altitude balloons providing a 1 Mbps signal to anyone within range on the ground. Loon is designed to provide low-end broadband in hard-to-reach regions where reliable service has been difficult or even impossible. As of 2020, the project seems to be in a late experimental stage.

Broadband over power line (BPL) can take advantage of all the electrical grid that connects power authorities with homes and businesses to provide internet data. Ultimately, the technology delivers limited bandwidth because line noise causes significant data signal lose. Data-carrying power lines can also cause interference with high frequency radio communications. In the end, relatively low signal quality and strong competition from other technologies mean that BPL will probably never be widely adopted.

Networks using forms of "wireless internet service provider" (WISP) can service homes and offices across larger geographic areas without the need to physically cable to every building. Wired connections are installed in an area's center and, where necessary, connected backhauls are installed in elevated areas to strengthen the wireless signals aimed at consumers. Existing radio towers or other tall structures can be used for the backhaul repeaters, making a WISP system relatively inexpensive to install.

Smaller-scaled wireless network co-ops can be shared locally using a "neighborhood internet service provider" (NISP) (using rooftop antennas) or a wireless mesh network (where network-connected devices act as peer nodes) to efficiently share a single physical connection.

Those systems are primarily designed to serve us where we live and work. But mobile data access is definitely also a thing. I'm sure you're already familiar with data plans that mobile phone companies can provide alongside their calling and texting services.

Mobile phone data access

Cell connectivity is distributed through geographic areas (known as "cells") from individual radio transmitters spread throughout the cell. Since the transmitters within each cell will use different radio frequencies than those in the cells around it, modern wireless technologies much implement a seamless, automated "handover" as a user moves between cells.

The technologies used for wireless telephony have changed since the 80s, when what's now known as 1G ("First Generation") cell phones were introduced. To describe the evolution of cell phones in very general terms, we could say that:

- **1G** phones carried only voice communications and had a maximum transfer speed of 2.4 Kbps.
- **2G** phones could carry Short Message Service (SMS) and Multimedia Messaging Service (MMS) messaging, which could include short videos and images.
- **3G** phones had much higher transfer rates (as high as 2 Mbps) than any variant of 2G and was therefore dubbed, "mobile broadband."
- **4G** phones could reach speeds as high as 100 Mbps, which permitted HD mobile TV, online gaming, and video conferencing.
- **5G** phones - when used on compatible networks - are expected to reach transfer speeds of up to 20 Gbps at a very low latency, permitting fully immersive virtual environments. Should the 5G rollout be successful

(and, at the time of writing, this isn't yet clear), the range and limits of new service categories that could be deployed are not yet known.

When it comes to planning a new venture, it's long been the accepted wisdom that there's no replacement for solid market research. Without knowing who your customers will be, where they live, and what they like, how can you properly serve them? Well, now you can add to that list "how reliable and robust is their internet connectivity," because without digital access, they may never find you or be able to consume your service.

Talking to the internet of things

Two recent changes are, more than anything else, responsible for the internet of things (IoT) ecosystem: the availability of cheap, embedded, single-board computers (like the Raspberry Pi shown in figure 4.1) and cheap and always-on internet connectivity.

Figure 4.1: The Raspberry Pi 4 Model B single-board mini computer

Those tiny single-boards - often smaller than a credit card - are easy to incorporate into just about anything you're planning to manufacture. Such devices cost very little - sometimes just a few dollars a piece - and they're generally built to run fully-powered (and free) Linux distributions. And network availability means that the vast streams of data generated by all those on-board cameras, sensors, and other peripherals, can be automatically sent back "home" for processing and analysis.

The dream of IoT

Here are some ways that IoT applications are already actively changing business and consumer practices:

- **Inventory control**. The very first IoT device was - arguably at least - a Coca-Cola vending machine at Carnegie Mellon University. Back in the early 80s, the machine was modified to digitally report its ongoing inventory. The simple idea that physical devices can monitor themselves and their surroundings, providing accurate, up-to-the-minute status reports to remote servers lies at the heart of countless modern industrial solutions. Modern retail, wholesale, logistics, and manufacturing operations now have constant access to their inventories, allowing them to understand trends and anticipate problems.
- **Agriculture**. Increasingly, modern farming incorporates robotic irrigation, fertilization, planting, and even harvesting technologies. All those robots running around your property are generating data and, from time to time, getting themselves into trouble. Moving that data "back" to administration servers is critical for keeping track of what's going, what might need fixing, and how your actual farm is performing. You can, therefore, expect that each of those devices will be part of someone's IoT.
- **Military**. Communication is key for military operations. But if even weapons, vehicles, and other equipment can communicate autonomously, and if there are servers dedicated to interpreting and acting on that communication, then you're already way ahead of the game. Sensors connected to each of hundreds of components for, say, a fighter jet, can

contribute to giving planners an unprecedented view of what's really going on.

- **Smart cities**. When information from sensors embedded in buildings, roads, public lighting, smartphones, and electrical systems are combined with data coming from traffic cameras and public agencies, the potential for data-driven insights is huge. Properly understood data can help cities manage their resources, utilities, and even traffic more efficiently, and better maintain their physical infrastructure.

- **Smart homes.** On a far smaller scale than smart cities, home appliances can be connected, monitored, and controlled through smartphone apps or remote servers. Smart home devices already include heating and cooling systems, light bulbs, robotic vacuum cleaners, garage doors, and security systems. Smart home devices can be controlled through phone apps but, in many cases, also through voice controlled devices like Amazon Echo (Alexa).

Conversations about IoT are always just one step away from *buzzwordism* - where words lose meaning and exaggeration becomes an acceptable lifestyle choice. Not all IoT stuff is actually IoT. Or, to put it another way, not all IoT is worth talking about. But here's one good way to categorize a particular technology: if, hour after hour, something generates more data than any human being could possibly absorb, then it's probably an IoT device.

Effectively dealing with all that data can be a problem. And that's not the only potential for trouble in IoT land.

The nightmare of IoT

In the information technology world, as a general rule, the more active network connections you have in your infrastructure, the greater your risk of being successfully attacked. That's because successful hacker intrusions usually come through badly configured or unpatched devices. The more public-facing devices you're exposing, the greater the chance one of them will have a serious vulnerability.

What kind of vulnerabilities are we talking about? Well, the US government's Common Vulnerabilities and Exposures database contains nearly 140,000 individual entries, each one representing a unique software weakness that could allow unauthorized access to and destruction of an IT system. There are threats impacting all operating systems (Windows, Linux, macOS), all formats (server, PC, smartphone), and all ages (there are active threats going back to the 1990s). And many hundreds of new entries are added each month.

In that sense, IoT devices are no different than any other kind of computer. But there is one way that they're a whole lot worse. Because you usually don't directly interact with IoT devices on an OS level, and because they're often commodity items that are purchased and deployed by the dozens or thousands, you don't instinctively treat them like computers.

Most of us, as an example, are aware that we should create complex and unique passwords for our laptops and WiFi routers. But your fridge? Just plug it in and it'll be fine! The problem is that many IoT devices - like "smart" fridges - have their own embedded operating systems and, usually, their own network interfaces.

There's a good chance that anyone driving down your quiet residential street can scan for available networks, quickly identify the brand of IoT device you're using, assume that you haven't changed the authentication credentials from their factory defaults, and log in to your private network. What makes things much worse is that many device manufacturers are still shipping their products with authentication credentials using some variation of admin/admin.

That's a big problem.

Leveraging federated identities

All this talk about the *dangers* presented by authentication and credentials should make you curious about how they can be used to generate some *good* connectivity stuff. In a single word, that'd be *federation*.

Identity federation is a technology for linking a single person's identity across multiple network services. Federation is what lets you log in to online gaming or web application sites using, say, your Google account credentials.

The upside of federation is that a single login can be all you'll need as you move between many of the online services you regularly use. That lets you reduce the risk of exposing your password through a vulnerable website. Of course, it also increases the damage that can come from a serious data breach of the servers used by your federation provider.

Federation can be used to integrate with third party single sign-on (SSO) authentication systems, like Kerberos, the Lightweight Directory Access Protocol (LDAP), and Microsoft's Active Directory (AD). When used with cloud services, SSO systems can securely permit automated as-needed access to private account resources for consumers or processes.

Besides convenience, all this authentication goodness drives opportunities for secure remote collaboration on large, complex projects - itself a fast-growing trend.

5

Understanding the Business of Technology Research

Getting a new technology out to consumers will usually require good people and boat loads of resources - including money. Lots of money. Much of that money will be spent on research and, more often than not, the hard research needed to translate a great idea into a usable product will be performed by someone whose job title isn't "entrepreneur." Sometimes, in fact, the research will be done by individuals who are barely aware that their innovations have any commercial value at all.

If you're here because you want to get the jump on cutting edge technologies, then you may want to keep an eye on the organizations that are known to produce practical research. Knowing who's big in research, who's funding it, and where the big bucks are being spent can give you useful insights into what might be coming next. From there, you're just a step away from, say, spending time learning the tools that'll come with the new tech or positioning yourself to profit when it finally shows up.

Who funds commercial science and why?

Once upon a time, major breakthroughs in serious scientific research were the products of private patronages. The Italian Medici family, for instance, famously supported many individuals whose work would prove pivotal, including Leonardo da Vinci and Galileo. However, the years leading up to the Second World War saw the scope and complexity of research projects growing far beyond the capacity of private support. The war's dependence on unprecedented technological complexity - exemplified by the work of the Manhattan Project building the atom bomb - pushed more and more research under government charge.

Government involvement in research has continued in the generations since the war. Still, it's been estimated that universities and governments are responsible for only 30% of research funding between them, with most of the rest provided by private industry (see en.wikipedia.org/wiki/Funding_of_science).

Let's see how that breaks down.

Taxpayers

Democratic governments, of course, don't spend their own money, of which they traditionally have none. Their many programs and services are funded by revenues raised, one way or another, from their capital assets and from their populations. In modern nation states, "populations" would mean those individuals and corporations who pay taxes.

Public research and development can be performed within government agencies. According to the terms of some agency mandates, research results must immediately enter the public domain. But even those who retain rights to their research will often point their work towards businesses and institutions that can use it productively. The US National Science Foundation (NSF), for instance, uses its $8 billion annual budget to fund "approximately 25 percent of all federally supported basic research conducted by America's colleges and universities" (https://www.nsf.gov/about/).

Other American agencies do much or all of their research in-house. Here are some examples:

- The *National Institute of Standards and Technology (NIST)* has a mandate to "promote innovation and industrial competitiveness." One very important part of that mission is maintaining the National Vulnerability Database (NVD) which plays a foundational role in the management of the vulnerability assessment and detection systems protecting our IT infrastructure.
- The US military's Defense Advanced Research Projects Agency (DARPA) collaborates with private and public sector partners to aid in the development of emerging technologies. Work in recent years has included research into robotics and autonomous vehicles, but you might be more familiar with a DARPA innovation from a few decades ago: the internet.
- The National Institutes of Health (NIH) employs 6,000 research scientists across 27 research institutes and centers. Their "mission is to seek fundamental knowledge about the nature and behavior of living systems and the application of that knowledge to enhance health, lengthen life, and reduce illness and disability."

The complete list of US government research agencies (available at en.wikipedia.org/wiki/List_of_United_States_research_and_development_agencies) makes for quite a read. Take a look for yourself.

Naturally, governments of other countries have their own research agencies. One example is Canada's National Research Council (NRC), which has evolved from its military technology origins through the two world wars, to its current focus on partnerships with private and public-sector technology companies. The NRC now divides its work into four "business lines:"

- Strategic research and development
- Technical services
- Management of science and technology infrastructure

- NRC-Industrial Research Assistance Program (IRAP)

As we mentioned when discussing the NSF, a significant proportion of taxpayer funds directed towards research and development are granted to public and private colleges and universities. But, from the college perspective, how much academic R&D funding comes from government sources?

A 2016 review of the 20 US colleges that spent the most on R&D found that they each spent between $837 thousand and $2.4 million and that between approximately 47-87% of their total spending came from government sources of one sort or another. (see bestcolleges.com/features/colleges-with-highest-research-and-development-expenditures/) By contrast, businesses only provided between 2 and 22% of that funding.

Private charitable funding

While we're on the subject of academic research, we shouldn't ignore a third source of funding: private endowments. Some - although not all - permanent endowments were targeted by their donors at research activities. Although the fund capital can't be spent each year, the income that capital generates can. Harvard university famously - or perhaps infamously - has a total endowment greater than 40 billion dollars. Some of that undoubtedly finds its way to R&D.

Curiously, according to that 2016 study, Harvard's total R&D spending that year - including activities funded by governments (52.1%), businesses (4.7%), and endowments - was just over one million dollars.

Of course, donations support plenty of research outside of academic settings, too. Most serious diseases have associated charitable foundations that exist to raise money for both victim care and medical research. And many thousands of registered non-profits exist throughout the world supporting non-medical causes, including many involving technology-related research. The Bill & Melinda Gates Foundation is a particularly well-known example.

Corporations

Technology-oriented companies have a strong interest in getting their hands on innovations before their competition. To improve their chances, many will run their own research labs in-house. For instance, the Bell Telephone Company - and its successors including American Telephone & Telegraph Company (AT&T) - maintained the active and enormously creative Bell Labs. Bell Labs, under various names, was responsible for many innovations, including the transistor, lasers, and the Unix operating system.

Individual technologists at some companies are often sources of innovation. 3M, for instance, has what they call a "15% Culture," where employees are allowed to use company time and space to pursue research based on their own ideas and interests. Over the years, the program has generated successful products for the company, including their sticky paper Post-its. In another example, Percy Spencer, working on radar for US defense contractor Raytheon, accidentally discovered that microwaves could cook food.

It should be noted that not all corporate innovation is truly home-grown. A lot of it is actually funded indirectly through government money in the form of tax incentives or credits. Under such programs, companies may be permitted to use research-related spending (including salary expenses) to reduce the income taxes they would otherwise pay.

Major fields of commercial technology research

Trying to grasp the full scope of technology development at this point in history is an unforgivable waste of time. There's serious innovation going on every minute of the day, in every time zone, through countless labs, office towers, warehouses, garages, basements, bedrooms and, of course, invisibly within creative people's minds. No one's keeping track of it all because it's not possible. Not to mention the fact that much of that innovation happens under a thick shroud of secrecy.

But it's probably worth offering just a couple of examples to give you a feel for where to look.

Quantum computing (and why we should care)

A cousin of mine with an advanced physics degree from Cambridge University once tried to explain quantum mechanics to me. He failed. Miserably. My poor old brain just couldn't absorb it. So don't expect any full, measured descriptions of the underlying science here. Instead, I'll try to show you how experimental *compute* technologies that depend on the physics might work, and what can be done with them.

The super-quick executive summary version of this is that computers powered by one quantum technology or other will work *a lot* faster than any of the super-est of super computers we have now. So much faster, in fact, that they may be able to solve problems that would be simply unfeasible using traditional computers (an achievement known as *quantum supremacy*). This would mean that some long-held assumptions about the way software works will no longer apply.

For instance, the reason the best encryption tools we currently use to protect sensitive data work, is because it would take hundreds or even thousands of hours of high-performance compute time to successfully break the encryption key. In most cases, it's just not worth the effort and expense.

But if you could easily buy time on a computer that processed operations exponentially faster, then two things would immediately happen:

- Cracking encryption algorithms would become trivial
- Honest folk would have to look seriously for a new way to protect their data

Currently, Google and IBM are among the major companies that have invested heavily in quantum compute research projects.

As well as I can understand it, quantum computers would measure the state of subatomic particles and use that binary measurement to represent a computational value. The description of that state is known as a *qubit*, which is effectively the quantum equivalent of traditional computing's *bit*. But because a qubit can also exist within what's known as *coherent superposition* -

meaning that its value can exist in a *superposition* of two possible states - it can be used to represent a more complex range of values.

And *that*, I'm given to believe, means that such computers will be able to do stuff much, much faster than they can now. If this actually happens, it'll be big and scary.

Energy technologies

The modern world consumes an awful lot of energy. We're constantly moving about, controlling our indoor (and in-transit) climate conditions, exchanging information, and expecting that all the world's riches be delivered to our doorsteps. By tomorrow.

But those energy-thirsty activities come with costs, not the least of which from the emissions they leave behind. The search for reliable, steady, and affordable energy sources that can help us find a healthy balance between consumption and emissions is ongoing; and unimaginably expensive.

Small modular nuclear reactors (SMRs) have been the focus of some serious developments in recent years. They appear to promise reliable, steady, and affordable energy in ways that their expensive and complex nuclear predecessors couldn't. First and second generation reactors were, overall, reliable and steady - and they were clean - but their massive capital costs and large physical footprints made them more than a bit inflexible.

The idea behind SMRs is that highly efficient reactors can be manufactured off-site and delivered on trucks one module at a time for on-site assembly. The design makes the per megawatt generation of power far cheaper and project completion much faster. And it allows the deployment of nuclear power to service smaller markets that previously couldn't consider it as a realistic option.

As the name implies, SMRs are smaller than traditional reactors. They're designed to deliver between 50 and 300 MW of electricity each, compared with the 800 to 1,200 MW outputs that were previously common.

Companies heavily involved in this research include Britain's Rolls-Royce and an American company with historical connections to the US Department of Energy called NuScale Power. Various governments around the world have also invested in the technology one way or another.

Medical technology research

If you think we're spending a lot of money on energy, wait 'till you see how much health care costs. Across the 37 Organisation for Economic Co-operation and Development (OECD) nations, health care industry spending accounts for around 10% of the total gross domestic product. That's more than $3,000 a year for every single man, woman, and child.

On the one hand, with all that money being thrown around, there are undoubtedly many business opportunities waiting to be discovered. But there's also a lot of room for new and innovative technologies that can improve the delivery of health care while reducing the costs. Here are two excellent candidates:

- Telehealth involves the provision of health services (like patient-doctor consultations) through a telecommunication medium. This might mean having a simple telephone conversation rather than a visit to the office, but it could also incorporate video conferencing tools or even the use of remote diagnostic equipment. For example, small, remote communities could maintain imaging facilities and technicians even many hundreds of miles away from the nearest medical labs and radiology specialists. Digital connections can permit distant doctors to view, say, ultrasound results, speak directly with patients, and confidently reach diagnoses. And all without the need for anyone to undertake exhausting and expensive travel. Telehealth also allows for meaningful patient-doctor contact without the risk of spreading disease.
- Telesurgery is an extension of telehealth which can allow some surgical procedures even when doctors are many miles away from their patients. The technology makes use of high-definition video feeds and purpose-built robotic arms that can be controlled by doctors remotely. Telesurgery tools have the potential to save money for cash-strapped health systems

but, more importantly, they can improve health care and save lives.

6

Where Hot Trends Go to Die

You're reading this book, so I'll assume you have an active interest in learning about hot technology trends. But you don't want to take every last one of these gadgets and business fads too seriously: some are bound to disappoint. (Although don't think you can hold me personally responsible for any of my predictions: by the time this book hits the streets I expect to be living on an unnamed sun-drenched tropical island under the witness protection plan.)

To get a sense of how fragile the innovation business is, keep in mind the popular wisdom that teaches us how nine out of ten startups will fail. Now multiply that by the particularly speculative nature of the technology industry in particular and you'll appreciate how easily things can often go spectacularly wrong.

That's not to say that the folks who dreamed up all the doomed businesses we'll soon discuss were fools or frauds. It's easy for us, enjoying the benefits of historical hindsight, to judge their efforts, and we should be sensitive to how different things must have looked in the heat of the moment. Still, bearing that in mind, there's value to be had from trying to at least understand what went wrong.

So here are some particularly impressive examples from the junkyard of tech history. It can be loads of fun to relive some of the biggest business disasters in history, but there are also important lessons we can apply when assessing this year's crop of "can't-fail" devices.

Market research beats wishful thinking every time

I'm not sure it's possible to reliably count all the individuals and companies that have assured us we've finally reached the age of bypassing choked highways using flying cars while happily chatting with loved ones through video calls. There have been hybrid wheeled/winged prototypes since soon after the Second World War, and telephones for showing the world how you look in pajamas have, in theory at least, been available since the early 1970's.

Well, video calls are now easily available through any smart phone or PC. And, 75 years of failure later, the rush to deliver consumer flight hasn't slowed down a bit. But both technologies have been most notable for being only rarely used: the cars because none has ever hit full production, and the phones because very few people seem interested.

What's been the trouble? There were certainly engineering, safety, and regulatory hiccups over the years. And there's no question that flying car manufacturers would be hard pressed to find a large customer base of drivers who were also qualified pilots (although self-driving/flying versions could, in theory, avoid that issue). But I suspect a big part of the problem was market research: no one bothered asking Joe Q. Customer for his thoughts on the matter.

But marketing isn't everything

In the beginning, there were comments from tech insiders about how this would be the biggest thing since ever. Then came an unauthorized book leaking intriguing information, some ambitious public claims, and a product launch. In the end, there was Segway: a personal transportation device that was too big and fast for sidewalks, too big and slow for roads, and too expensive for most customers. And using it in the rain or snow wasn't much fun at all.

Today you'd probably have to look pretty hard to find a living, breathing Segway anywhere close to your neighborhood. They're sometimes used for police street patrols and touring, but they haven't eliminated the car or revolutionized urban development. Nor, as far as I can see, did they make the company's investors fantastically wealthy. In fact, the company's manufacturing plant in Bedford, New Hampshire ceased operations in the summer of 2020.

What went wrong? Well, perhaps the hype was a bit over the top. Ok, make that way over the top. It's never a good thing to pump up expectations to the point they can't possibly be met. There was also the failure to match the tool to an appropriate environment. Where, after all, was it supposed to be used? But, to be really successful, a new product has to be built on more than clever engineering. It also has to solve a real and pressing problem.

When too much power isn't a good thing

Back in 2013, Google introduced a new consumer compute product they called *Glass*. This was a sleek headset that could be worn as an attachment to a pair of designer prescription glasses. When powered on, Glass could accept voice and touch commands to record video of everything the wearer sees, and display data - often with full "awareness" of the wearer's current physical location.

Glass was a single device intended to replace much of the function currently served by smart phones, laptops, and media players. For the task of integrating our physical world with the endless data that describes it, this was going to be perfect. And then it wasn't.

As more details about Glass became known, questions were raised in the broader tech world. Was it appropriate - or even legal - to silently record videos of other people? Should face recognition software be applied to random pedestrians walking past on the sidewalk without their consent? Was it safe to drive while wearing Glass?

Potential customers had their own questions. Is the product affordable (they started at $1,500)? Is it necessary? Does it fit the vision I have for my public image?

The longer those questions floated around the internet, the more answers came back. Answers, by and large, consisting of a single word: "No." Google

Glass, as a consumer product, slowly faded away and eventually disappeared altogether. The massive media promotion campaign had come up empty.

Which is not to say that the product itself failed. As it turns out, Glass has found considerable success in medical environments where, for instance, it could be used to permit remote surgical experiences. It's also found a home in industrial settings, where front line workers often need instant, hands-free access to relevant schematics and directions.

But it was a long while before all that goodness happened. Perhaps someone should have slowed things down at some point, saying: Even if it's *possible* to engineer all of those features into a consumer product, is it necessarily a good idea?

When a thousand pieces don't all fall where they're supposed to

Sometimes measuring success and failure isn't so easy. Take WebTV as a case in point.

Who doesn't own a TV? (Besides me, I mean.) Wouldn't it make sense to create an inexpensive and easy-to-use product that leverages billions of existing home TVs for non-standard but popular uses? How about a device that can turn the TV you already own into a web browser and email portal? If that doesn't sound so exciting today, back in the mid-90's the idea behind WebTV had its definite charms.

Just imagine the secondary revenue streams this could generate. Wouldn't advertisers climb over each other to pay big bucks to have their products pitched to all those TVs?

Had WebTV managed to deliver on the "easy-to-use" angle, things might have gone differently. But it turned out that the primary demographic for the device was heavily skewed to older people; who needed a lot of (expensive) customer support coaching through the setup process. The company's inability to keep up with the fast-changing internet browsing standards also made it tough to provide a consistently optimal browsing experience - especially for users sitting ten feet away from the screen on their couches.

How did things actually play out? On the one hand, within two years of their launch, the company was purchased by Microsoft - who rebranded the service "MSN TV" - for more than 400 million dollars. In one form or another, they stuck around until long past the death of dial up internet access. So that's a good thing.

But, arguably, they failed to capture nearly as much interest and adoption as they could have. The real prize was in becoming a dominant portal for internet access. Because the platform was proprietary, the company could effectively have controlled the entire internet experience of hundreds of millions of users. The potential scope of the product would have dwarfed the modest revenues they actually achieved. So that's not a good thing.

Were they too far ahead of their time? Did they miscalculate by insisting on a closed, proprietary platform? Did they fail to see the monstrous growth in the standalone personal computer (PC) industry coming?

Either way, unless you're the type of person who cheers for the evil stepmothers and hungry wolves, it wasn't exactly a fairy tale ending.

When timing isn't your thing

The tech industry moves fast. I'm sure that little nugget of wisdom won't leave any of you wrapped in stunned silence. But when you think about how much work is needed before you can convert a fresh, new idea into a ready-to-ship product, it's remarkable anything innovative ever gets off the ground.

Bad timing, then, is a risk faced by the people behind pretty much any new technology as it makes its way to market. By way of example, the existence of strong competition from companies like Nintendo and Sony's PlayStation were probably largely to blame for the premature death of Apple's Bandai Pippin gaming console back in the mid nineties. Although, the fact that, at peak, there were never more than 25 game titles that would run on the device and that, like all Apple products, it was priced much higher than the competition, couldn't have helped.

All wasn't dark and foreboding for Apple in those years. Looking back with what we now know, the strong presence of their iPod digital music player platform was probably what doomed Microsoft's Zune. On that one, Microsoft had the bad luck (or lack of foresight) to get stuck between an iPod device made dominant by its simplicity, and the looming age of the smartphone (which made standalone portable music players irrelevant).

Clearly, as Shakespeare would have it, "ripeness is all."

But there's another thing about timing: eventually, you'll need to deliver the goods. There's a limit to how long we'll wait for that bright new technology that's been on everyone's "must-have" list for too long without making an actual, real-world appearance. Beware empty promises.

You should also keep a critical eye out for good old fashioned bad business practices - the kind that never seem to go out of style. I'm thinking about unrealistic business plans; unfamiliarity with a business' core, underlying fundamentals; and unreasonable, greedy start up costs.

Actually, I'm thinking about the catastrophic disaster that characterized the dot-com boom and subsequent bust around the start of the 21st Century. The take-any-business-model-and-build-it-a-website paradigm looked good, but it was applied far too broadly and often ignored the obvious context in the process.

Don't blindly trust popular trends and buzz phrases.

II

Buzzword Bingo: Quick and Dirty Technology Profiles

The first part of this book covered some "big-picture" technology topics like privacy and connectivity, following their twisting tentacles wherever they'd lead us. In this second part, we're going to focus on a couple dozen specific technologies. These technologies have names that you've probably encountered and that, for the most part, already play important economic roles. Sometimes their significance and impact might be exaggerated, but they're all the real deal.

7

Compute Platforms

The way things sit now, if you were somehow allergic to computers, you'd be hard pressed to really banish them from your life, no matter where you found yourself. Taking a quiet walk in the woods? What about the smartphone in your pocket. Left the phone in the car? See that cell phone tower just behind those trees? The odds are good that the tower is more than just an antenna; it could also be hosting an edge computing server. And don't think that there weren't computers embedded into the under-the-hood workings of the car (or bus) that drove you over.

Allergies aside, if you want to fully grasp the current state of the compute world, it'll be helpful to understand all the places computers can pop up and what they might look like. In this chapter, we'll enumerate the classes into which modern compute devices can fall, and describe their strengths, weaknesses, and potential.

What is a server?

Honestly, I'd been working as a professional system administrator for a while before I could have properly answered that question. The truth is that every server is a computer, and any computer can be a server. The term *server* simply implies that the device is providing some *service* to at least one external device (known as a *client*).

If the printer that's plugged into your desktop computer can be shared by the other computers in your local network then your desktop is a server (a *printer server*, to be precise). The WiFi router provided by your internet service provider is, by all definitions, a *network server* - as it *serves* network access to its clients. And the tiny, $5 Raspberry Pi Zero single board device (as shown in figure 7.1) that powers your homemade surveillance camera is a *video server* - although that won't work without attaching a $7 camera module.

Figure 7.1: The Raspberry Pi Zero - a fully-functioning server for under $10

But that's not what most people are thinking about when they use the term. The first time I ever walked into the server room belonging to a mid-sized business, I was hit by the sound of dozens of powerful chassis fans and the heat from hard-working CPUs and fast-spinning disk drives. Instantly, I knew how folks normally use the word. (I also soon discovered that the heat was a problem: the admins were struggling to keep their server room properly cooled and would, over time, end up having to write off some expensive hardware due to heat-related failures.)

So, by "server," we usually mean computers installed within those rack-mounted, stackable cases built to efficiently house and protect highly performant, expensive, and delicate components. Server racks will normally live in well vented and cooled rooms with easy access to ample electrical power. You may have to search for them, but such rooms will also always contain colorful bundles of cabling, connecting the servers to networks.

As a rule, servers won't usually have displays or even keyboards plugged in, as they're likely to be managed remotely or, even more likely, fully automated and requiring no administration at all.

Server farms belonging to giant cloud providers like Amazon Web Services will have many thousands of commodity computers running within aisle after aisle of vast warehouses. When one fails, a monitoring panel somewhere will light up and a technician will eventually be dispatched to remove the server, throw it out, and slide a replacement into the newly-available rack.

No tears are shed when we say goodbye to hardworking and devoted old hardware in those places.

What is Linux?

Speaking of servers, did I mention that they all need some kind of operating system installed? And did I mention further that the vast majority of the servers powering the vast majority of the operations that make the internet and all its functionality possible are running the Linux operating system? Oh, and did you know that the open source Linux operating system is available for free?

I didn't mention all that? My bad.

Well, servers need operating systems. Most servers (well over 90 percent of the virtual machine instances running on Amazon's AWS EC2, for instance) run Linux. And Linux is, indeed, freely available for any use on any server, laptop, desktop, router, embedded system, or supercomputer. In fact every last one of the world's top 100 supercomputers uses Linux. And the Android smartphone OS? Yup. It's Linux, too.

Strictly speaking, "Linux" is the software kernel that allows a computer user to take control of a computer's physical hardware elements. The kernel translates your keystrokes and mouse clicks into a format that will be understood by the drivers controlling your storage drives, memory, network interfaces, displays, and - in fact - keyboard and mouse. Many thousands of additional software programs are closely associated with Linux, but they're actually part of the user space that hovers "above" the Linux kernel.

Having said all that, Linux, including its broader software ecosystem, dominates the server computing market right now. The fact that you can freely install and fire up as many physical or virtual instances as you like makes Linux very attractive, especially in the world of scripted workload orchestration. Virtualized Linux instances will often be brought to life and then, after completing a task that takes even a few seconds, killed off again. The versatility and flexibility Linux brings to computing have been the spark of some deeply impressive innovation and creativity.

Part of the Linux versatility is the fact that you can choose from among hundreds of variations (known as *distributions*). Are you looking to run enterprise supported servers? Internet of Things (IoT) devices? Security testing machines? Multimedia management? Video or audio production? All of the above? None of the above? There's bound to be a distribution that's a good match for you. And if the exact specs you need can't be found, feel free to rewrite the kernel itself and create your own distro.

Full disclosure: I know a thing or three about Linux, being the author of Linux in Action (Manning), a coauthor of Ubuntu Bible (Wiley), and the author of the Linux Fundamentals learning path at Pluralsight.

Fuller disclosure: I'm writing this on an Ubuntu Linux workstation in my home, where all of our many devices have been Linux-powered for more than a decade.

What is virtualization?

We've already discussed virtualization in some depth as part of Chapter 3 (Understanding the Cloud), so we'll just cover some big-picture conceptual basics here.

In the old days, when you'd get an idea for a new compute project, you'd need to:

- Submit a proposal asking for money from your managers
- Estimate your requirements
- Solicit bids from hardware vendors
- Place an order for a new server
- Wait for it to arrive
- Wait some more
- It arrived! Unpack!
- Install your application software
- Launch your new server

That's the way things used to work: One project. One server. Lots of waiting time.

But what if you overestimated your compute needs by 50 percent? That'd be a few thousand dollars down the drain. And if an important but lightweight project didn't really need a full, standalone server, you'd often have to buy it anyway. How about if the project would only have to run for a few months? Spend the money and hope you'll find a new use for the thing once your initial project shut down.

Awkward. Mountains of awkward.

Virtualization is a (mostly) software trick that lets you fool multiple installed operating systems into thinking they're all alone on a physical computer when they're really sharing it with other OS's. You can provision and run a single virtualization *host* of one flavor or another and then fill it up with one or a hundred virtual servers.

One of those servers might need a lot of system memory but only a GB or two of storage space. Another one might be heavy on video conversion tasks and storage but is only needed for a half an hour a day. A third could be a 24/7 monitoring system that just needs a quiet place to do its thing without anyone bothering it.

As long as you never push the physical host past its overall resource limits, the virtual machines can all coexist happily together. And when one service is no longer needed, you can reassign its freed-up resources to something else.

The ramp-ups and ramp-downs of a typical virtual server's life cycle are fast. For all intents and purposes they'll generally launch and shut down instantly. This is possible because the underlying hardware is always running - and because the OS image is small and, usually, optimized for virtual environments.

As we saw back in Chapter 3, cloud-hosted services are all virtualized. As more and more IT infrastructure moves to the cloud, more and more of your online activities will be driven by virtual machines. You won't notice the difference, but every time you search the internet or authenticate to an online account, there's a good chance that it's a container or VM you're connecting to, and not directly to a physical machine.

Where do you go to get some o' that there cloud?

Like virtualization, we also talked about the cloud back in chapter 3 - which would make sense, considering that the chapter was called "Understanding the Cloud". We mentioned how the public cloud market was dominated by AWS and, to a lesser extent, by Microsoft's Azure. I'll just take a minute or two here to add a quick guide through some of the cloud industry's worst jargon.

- **Infrastructure as a Service (IaaS)** environments give you full access to virtual server instances. The provider will ensure the underlying hardware, networking, and security elements are in place and functioning, while it's your job to manage the OS and other software running on your instance. Major IaaS players include Amazon's Elastic Compute Cloud (EC2) and Azure's Compute.

- **Platform as a Service (PaaS)** environments hide most or all of the infrastructure administration tasks from you, leaving you with an interface where you can run your own data or code. One good example is AWS Elastic Beanstalk, which lets you upload your application code from where it'll be automatically deployed to Amazon's cloud. Other providers in this space include Heroku and Salesforce Lightning Platform.

- **Software as a Service (SaaS)** environments expose only an end-user interface, managing all layers of the administration infrastructure invisibly. Microsoft's Office 365 and Google's G Suite are widely used SaaS office productivity tools. But there's a growing marketplace of SaaS tools providing online software equivalents to many applications that, in years past, could only be used on standalone workstations. Those applications include accounting, computer assisted design (CAD), and graphic design solutions.

- **Consumption-based pricing** or, as it's sometimes known, pay-as-you-go billing, is a cornerstone of the cloud concept. The idea is that you don't have to gamble by investing up-front in infrastructure, but you can pay incremental amounts for units of compute services as you use them. It might not always come out cheaper in the long-run, but pay-as-you-go definitely makes it easy to test application stacks and experiment with multiple alternative configurations before pulling the trigger on a full deployment. It also means that - assuming you don't make any dumb configuration mistakes - it's nearly impossible to badly over-provision.

- **On-demand** is also sometimes referred to as self-service. The ability to request instant delivery of compute resources any time of the day/week/year gives you complete control over your organization's application life cycles. You're never at the mercy of other people's schedules and limitations.

- **Service Level Agreements (SLAs)** are legal disclosures published by companies in the business of providing services. Even if the standard of resource reliability provided by the major public cloud platforms is generally excellent, accidents will happen. When you pay hourly or monthly fees for cloud services, the company's SLA tells you that you should anticipate downtime of a certain number of minutes or hours each month. As an example, Amazon's SLA sets its EC2 availability rate at 99.99% each month. If, in a particular month, you encounter greater downtime, you might be eligible for service credits or refunds.

- **Multitenancy** is the placement of virtual instances belonging to multiple cloud customer accounts on a single hardware resource. A multitenancy setup for a server instance will probably be significantly less expensive than a dedicated instance. Choosing a dedicated instance, however, would guaranty that your instance will never be hosted on a physical server alongside an instance from a second account. Security or regulatory considerations might require that you avoid multitenancy.

- **Migration** describes the process involved with moving existing business application and database workloads from local (on-premises) deployments to a cloud provider. Providers often make specialized tools and free tech support available for migrations.

- **Elasticity** describes the ways virtualized cloud resources can be quickly added to meet growing demand or, equally quickly, reduced in response to dropping demand. Elastic resources are especially well suited to maintaining application availability and health without incurring unnecessary costs. Elasticity can usually be automated, so applications will respond instantly to changing environments without the need for manual intervention.

What is "serverless" computing?

Serverless computing is no different from server computing. It's just that, even if you squint your eyes real tight, you don't get to see the server. Or, to put it another way, serverless computing is like running a virtual server instance, but without having to configure its instance settings or log in to set things up.

In other words, you can't run software code of any kind without a computer somewhere processing your commands. So let's say that "serverless" is a form of virtualization where everything except your application code is abstracted. In that sense, serverless platforms like Amazon's Lambda and Azure's Functions are a lot like the model used by Amazon Elastic Beanstalk, except that they're so simple to use that they can easily be incorporated into a larger, highly automated multi-tier workload.

What is Edge computing?

Latency is the term we use to describe the time it takes for data to travel from a remote server across a network to your computer - or back in the other direction. Assuming you prefer fast service over slow service (which seems a safe assumption), high latency numbers are a bad thing.

Network engineers can invoke various magical spells - Oops! I mean clever configuration efficiencies - to reduce delays due to latency. But no matter how many tricks they're hiding in their mysterious black bags, they can't ignore the laws of physics. Even using the best connections and configuration profiles, data still has to physically move across the distances between remote locations.

The only way to reduce this kind of latency is to shorten the distance. I suppose that one way would be for online service providers to very politely ask their customers to sell their houses and move somewhere closer to the servers running in the office (as if real estate prices weren't already high enough in Silicon Valley). Alternatively, though, how about moving the server closer to the customer?

Ah. You've discovered edge computing: the fine art of installing large distributed networks of smaller servers where mirrored copies of server data can be stored and, when necessary, fed to any customers in the area who initiate requests. If you've got enough of those servers spread evenly through the geographic regions where your customers live, then you can significantly reduce the latency they experience.

One kind of edge computing that performs this function is known as a content distribution network (CDN). Cloudflare and Amazon's CloudFront are among the larger CDNs currently in operation.

Edge computing resources like those used by CDNs have also increasingly been used to manage large streams of data from and to IoT devices - like the computers embedded in cars. Placing capable compute devices at the edges of large networks makes it possible to consume and transform such data sets faster than by moving the data all the way back to the more distant cloud.

What are the key compute form factors?

Computers, like egos, come in all shapes and sizes. Would you like to carry a rack full of bare metal servers around in your pocket to pay for your shopping? You're probably better off using some kind of mobile payment app on your smartphone.

Size matters. A lot. A device's form factor will determine the dimensions and capacity of its internal components. That means the particular motherboard, memory modules, storage drives, peripheral ports, and power supply you select for a device will be limited by your overall form factor.

The form factor you choose - for either a new project or just for your personal use - will generally be obvious (server racks can be heavy and don't handle travel well). But knowing what's available can make it easier to plan.

Devices using video displays

The term *personal computer* (PC), these days, is used to describe either desktop or laptop computers. Laptops, since they're designed to be mobile, are largely self-contained. Desktops, by contrast, generally come with the core compute elements within a box that includes external ports for connecting peripherals like keyboards and monitors. While you can find computers with power and functionality that's comparable to PCs in very small (credit-card sized) cases, the larger boxes used by desktops allow for easier customization and upgrades.

Gaming consoles - like Sony's PlayStation, Microsoft's Xbox, and the Nintendo Switch - are effectively the equivalent of desktop PCs, except that their software is built on closed systems. They're "closed" in the sense that their software interface exposes only the functionality the manufacturer wants you to see. Modifying or customizing the OS or internal works of a game console is normally impossible.

A touchscreen device uses the gestures and taps it senses from users as input devices in place of the traditional mouse or keyboard. Touchscreen technologies are the primary inspiration behind smaller form factors for consumers, since there's no need for external input devices. This, more than just about anything else, has driven the tremendous growth of the tablet and smartphone markets. (It also explains the freakishly agile thumbs of entire generations of young people.)

Devices without video displays

The router that connects devices to a network through either WiFi or ethernet cables contains pretty much the same internal motherboard and network interfaces that you'd find in any other compute device. The big difference is that there's no HDMI, DVI, or VGA video port. Routers are meant to run autonomously and, when administration is necessary, it'll usually happen through a browser interface across a network.

You launch an admin session with your router by entering its IP address into your browser and authenticating when prompted. In some cases, you can also launch terminal sessions through the Secure Shell (SSH) protocol.

This remote administration model is shared by many display-less device types. Those will include medical (and non-medical) implants or *wearables* that come with tiny computers built to monitor, report, or even interact with their host environments.

Go ahead and call me slow and old fashioned. But in the context of wearables, I would at least briefly discuss smart watches here. The problem is that, for the life of me, I can't figure out why anyone would want one.

Display-free computers are also embedded in medical devices, appliances, cars, logistics fleets and industrial machinery. All of those embedded computers are components of the growing internet of things.

8

Security and Privacy

Let's get some administrivia out of the way right off the top. You'll no doubt recall how security and privacy were covered like a blanket in the first two chapters (named "Understanding Digital Security" and "Understanding Digital Privacy" respectively). So why are we flogging this certifiably dead horse now?

Because it's not dead. Security and privacy are at least as important as anything else in IT. Even if they're things most of us don't think about enough, they're not something you can overdo. As an outstanding IT professional I once worked with would have said: "Paranoid is only the beginning."

And besides, there are still some urgent and fascinating topics we haven't addressed.

So we'll spend some time exploring how the core security tools (like authentication controls and encryption) can be applied to solve a much wider range of security and privacy problems. And we'll also go face to face with a couple of significant threats that exist thanks to the very devices we've come to love.

Blockchains

The new-technology-hype machine just loved blockchains when they first came to public attention. There were frequent gushing articles in the media about how this was *it*: blockchains were poised to change the world, ushering in a golden age of endless joy and fluffy fairy unicorns. Rejoice! Salvation is come.

But despite all that, blockchain technologies are, in fact, a big deal. Before we go there, though, just what is this stuff all about?

A blockchain is a distributed string of digital records used to record and validate transactions. The goal is to maintain a reliable and incorruptible public "ledger" of transactions to secure and improve the way financial and commodity operations are recorded.

The *blocks* in *blockchains* are actually data packets containing some identifying meta information (including a timestamp) and a cryptographic hash. The hash - which is produced by software running on the computer that generates the block - is derived from the unique contents of the previous block in the chain which, in turn, was based on the block that preceded it.

Because the contents of one block are dependent on the state of others, no single block can be modified without leaving behind some obvious and easily traceable evidence. This explains why it's called a *chain*, because if any one link (block) is altered, the entire chain will break. In effect, a chain will never be trusted unless it maintains the clear consensus of the creators of all its blocks.

Generating the hashes for blockchains is compute-intensive and can incur significant costs in compute power and electricity. This is by design, since it all but forces blockchains into the hands of distributed communities, rather than individuals or small groups. This decentralization makes chains less vulnerable to attack and adds robust reliability to the data that's being managed.

Blockchains and cryptocurrency

Like most people, I first heard about blockchains in the context of cyptocur-rencies like Bitcoin and Ethereum. Cryptocurrencies are digital assets that can be used as alternatives to fiat money. Fiat money, by the way, is the kind of virtual and mutually accepted representations of value found in exchange instruments like national currencies.

Using the funds in a cryptocurrency account, I could pay for goods or services while, in many cases, retaining anonymity. Of course, this very anonymity carries significant risks.

Cryptocurrencies have, for instance, been used to support criminal activi-ties. The people behind ransomware attacks will often demand cryptocur-rency payments in exchange for the decryption keys that you *hope* will restore access to your lost data. And the contents of large cryptocurrency accounts have been effectively lost when controlling servers crashed (or were forced down) or, in at least one case, when the administrator of a currency worth millions of dollars died without sharing his authentication information.

It's worth noting that the relative value of funds in the account itself - when measured against the ability to exchange them for fiat money - has historically been volatile, unpredictably suffering from violent market fluctuations.

Blockchains and accounting

Blockchains can solve many of the same old problems addressed by traditional accounting practices. Specifically, integrating blockchain verification into a business's financial processes can provide secure transactions and on-demand access to immutable and transparent records. The ongoing, real-time existence of such records could possibly remove the need for periodic audits and monthly reconciling.

Many of those same features could profoundly change the very nature and value of contracts - a change that could spill over beyond accounting, into the practice of law.

Blockchains and insurance

The potential security and privacy features of properly designed blockchains can also create efficiencies and value in the insurance industry. For one example, having a single blockchain where all the insurers within a particular market can reliably share their customer account information can help reduce claims fraud. Suspicious behavior and multiple claims for a single event will be more readily visible within a transparent and highly accessible system that includes data from all participating parties.

Being able to reduce administrative duplication can also greatly streamline the processing of legitimate claims. You'll appreciate this when you consider how a victim's insurer will often process their customer's claim using similar steps to those used by the insurer you're claiming from. But if both companies are able to openly share their data, the process can be unified and, even better, automated.

Perhaps most significantly, the delivery of healthcare can be enhanced and made more efficient if critical personal records can be safely and instantly accessed. And - you guessed it - blockchains can be helpful here, too.

What kinds of automation are we talking about? Well, going back to accident claims, a "smart contract" is software that regularly checks for changes to the status of associated objects. The simple mouse click approval of an insurance appraiser, for instance, could set into motion all the events necessary to pay a claim, notify all related parties, and update existing records.

Maybe - just maybe - insurance isn't as boring as people think.

Multi-factor authentication

Passwords are terrible things. Sure, we can't just leave our devices and online accounts open to anyone. But who decided that asking people to memorize long strings of meaningless text (like *sIIkdm^&sv234LKi*) was the solution?

Sure, you could choose easy-to-remember passwords like *mysecret* or this clever variation: *mysecret22*, but anything that's that easy to remember is equally easy to guess. And double that if you're using the same password for multiple accounts. In other words, that kind of protection is just not worth the effort.

There are, by the way, two ways to improve your passwords:

Use a password vault to generate and safely store insanely complex passwords that you won't need to remember: you can just copy and paste them into any login pages you visit. Use long (15-20 character) passwords that incorporate memorable, but unconnected, words. Something like:

house-seventy-warfare-calf.

Mathematically speaking, it's highly unlikely that anyone will have the compute power and time needed to crack that one. And it's not so hard to memorize.

But when it comes to particularly sensitive sites - like the ones where you do your banking - even good passwords aren't good enough. For that reason, more and more organizations are incorporating multi-factor authentication (MFA) into their security profiles.

A website or application configured with MFA requires you to present more than one kind of evidence that you are who you claim to be. One could be based on something you know, and another could be evidence based on something you have.

"Something you know" could be a password, while "something you have" could be a standalone MFA device or an app running on your smartphone. It'll often work by having the application send a short-lived code via instant message to a preset phone number. You'll be expected to enter the code onto the authentication login page.

Federated identities

Once you've got the basics of authentication out of the way, through strong passwords and/or MFA, there's the question of authorization. In other words, what resources your logged in account will be able to access. Individual systems will control users through some kind of access controls. Microsoft Windows, for example, uses Active Directory, Linux has object permissions, and cloud providers like Amazon Web Services can apply roles and policies.

But if you want your users to be able to move *between* services without having to log in to each service individually, or if you would just prefer not to have to manage authentication at all, you can implement a federated identity.

You've probably already experience federation without even knowing it. Logging into a third party web service using your Google account is one form of federation. The service integrates its authentication system with a federation provider using an identity technology like Security Assertion Markup Language (SAML) or OAuth. When you accept the terms and log in, the provider will share just enough of your identity information with the third party service to enable an account.

Digital surveillance

Because it can both protect you from harm and also invade your privacy, surveillance is a two-edged sword. But *digital* surveillance is a two-edged sword that's a whole lot sharper. Let me explain why that is.

Closed circuit video cameras have been in use within security systems since at least the 1930's, but they really did only one thing: record images that were usually stored locally and then, after a few days or so, overwritten with new recordings. That was helpful but, to make it useful, you would need to physically get to the tape and then laboriously search through, find, and view any images of interest.

Digital surveillance cameras are certainly cheaper than their analog equivalents, much easier to physically hide, and easy to access through networks. But there's also a lot more you can do with digital video feeds.

You can, for instance, configure email alerts whenever the camera detects motion. Or you can redirect a video stream to cloud services (like Amazon's Kinesis) where it can be integrated with your data analytics and machine learning operations or interpreted in near real-time by an object and face recognition service (like Amazon's Rekognition).

All of those tools can be used in the service of both positive and harmful goals. The fact is that there are now countless millions of such cameras deployed around the world that are, in many cases, connected to large-scale surveillance operations. At the very least you should be aware of the potential and risk such technologies present.

Backdoors

A *backdoor* is a hardware or software-based vulnerability that was intentionally built into a device or the operating system that runs it. In some cases, the backdoor exists with the full knowledge of the customer, as it was intended to enable remote support or the automated installation of patches and updates. But that's not always the case.

Governments, government-associated companies, and criminal organizations have been caught shipping sensitive compute and networking devices with dangerous backdoors. Such vulnerabilities have been used to bypass encryption protection to monitor communications, steal research data, and harvest authentication information.

Backdoors can take the form of active malware that collects local data and then sends it to remote attack servers. Alternatively, a backdoor can be designed to passively permit remote logins through insecure network environments.

Protecting yourself from backdoors requires defenses on multiple levels, including:

- Careful vetting of potential hardware vendors (taking into account their home countries and associations)
- Regular monitoring of reliable technology information sources for news of new vulnerability discoveries
- Careful monitoring of your devices' network activities
- Regular patching of your networking and compute systems
- Blind, stupid luck in large doses

9

Managing Data Storage

We've all been at this 21st Century thing for a while and by now it's pretty clear that data is the big driver of, well, of everything. Governments build their policies around economic and population data. Scientists build their theories around environmental, physical, and biological data. Businesses build their plans around production, sales, and consumer behavior data.

Data is being generated at rates previously undreamed of. I've read that the sensors on a pair of General Electric GEnx engines on a Boeing 787 Dreamliner generate a terabyte of data each day. A single network-connected car (like a Tesla) might upload around 100MB of location, performance, and maintenance-related data on any average day. Multiply that by the millions of such cars that will soon be in use around the world, and multiply *that* number by the thousands of other devices that are out there, and the scale of the data "problem" should be clear.

Got plans to add your own data to the flood and you feel the need to save and store it, too? You'll need to be able to explain why you need it so you'll know how it should be done. I can't help you with that "why," but I think I can give you some useful thoughts about the "how."

The *way* you store data will depend on what it looks like as it's produced and how you may need to access it later. *Where* you store your data will depend on how much of it there is, how deeply you'd be impacted by its loss, and how often you'll need to take it out and play with it. Let's take a look at both of those variables.

Data storage formats

Since not all data is created equal, it'll make sense to look for the tools and environments that'll most closely match the work you're planning to do. Here are some options:

Spreadsheets

They may be flashy, colorful, consumer-facing applications, but spreadsheets are no lightweights when it comes to serious data processing. As we'll see in more detail a bit later, spreadsheets have their limitations. But when it comes to presenting data in visually accessible ways, applying mathematical, statistical, and financial operations to that data, and even integrating remote data sources (like stock market quotes), no other tool comes close.

Spreadsheets can import simple, plain text data from files of just about any size as long as the text can be delimited. That is, breaks between data divisions should be marked by some consistent character.

When you import the data, you can specify the appropriate delimiter. Tabs, hard returns, and commas are common delimiting characters. In fact, the popular acronym *CSV* stands for *comma-separated values.* Here's what a few lines of CSV text might look like. Note that the first row contains column headings. Spreadsheets can easily understand how those should be treated differently.

```
Year,Volume,Rate,Growth
2015,56,10,15
2020,90,11,(2)
2022,109,8,12
```

Spreadsheets display their data in cells, which are arranged into horizontal rows and vertical columns. Functions can be applied to the contents of individual cells or to some or all of the cells in a column or row, and can incorporate values in cells in relative locations. Data sets within a spreadsheet can be rendered as graphs. Spreadsheets can be used as web forms where users can input data that's saved for future use.

The most popular spreadsheet is probably Microsoft's Excel, which is part of their Microsoft 365 Office Suite. But feature for feature, the open source Calc that comes with the LibreOffice suite is a viable alternative. Google Sheets is a cloud-based spreadsheet solution that may lack some of the feature depth of the others, but is a strong collaboration tool.

Databases

As a rule, databases are not built for visualizing data in attractive and intuitive formats. And, on their own, they're not known for complex mathematical calculations either. But boy, can they handle extra-large data sets and multi-table relationships.

When I say that databases don't really help you visualize your data, that's generally because they're meant to be used "behind" front end applications in multi-tier deployments.

For instance, an e-commerce website will display web pages where users can:

• Browse what you've got for sale
• Add items to a virtual shopping cart
• Check out using their payment information.

The web page itself just draws a graphical interface and shows you where to click your mouse, but the information about the items you're selling - including their price and associated images - are probably retrieved from a backend database whenever the page loads.

Similarly, your selections and, eventually, payment information will be written to a different database. The software process that handles your shipping might later consult the payment database for the shipping address. Databases are there at every stage, but no one will ever actually see them.

Administrating large databases so they're efficient, secure, and reliable takes serious engineering and, in some cases, an enormous amount of money. Before you build your database deployment, you'll need to know whether your operation requires strong Atomicity, Consistency, Isolation, and Durability (ACID) and support for complex and flexible queries. If it does, you may be looking for a relational database engine like SQL Server, MariaDB, or Amazon's Aurora.

Or perhaps you need speedy performance and would prefer a more flexible schemaless environment (suggesting you'd be better off with a NoSQL solution, like MongoDB or Redis).

SQL, by the way, stands for *structured query language* - which is an established syntax for using language-like code for interacting with your data. Counterintuitively, depending on who you ask, *NoSQL* might not stand for *Not SQL*. Some prefer to think of it as *Not Only SQL* instead.

Jupyter Notebook

Don't think you have to consume your data using the same tool that's storing it. It's possible, for example, to import existing data that's stored either locally or at a remote site into an interactive compute environment like a Jupyter Notebook. The advantage of this kind of setup is that the data can now be addressed within the context of, say, a robust Python programming environment without actually touching - or potentially corrupting - the original source.

The open source JupyterLab is a popular resource for working with large data sets using Python. You can download and build your own JupyterLab or run it remotely through a cloud provider like Amazon's Elastic Map Reduce service, or Microsoft's Azure Notebooks. For particularly large data sets - especially those that already live in the cloud - an existing cloud platform can make sense.

Data storage devices

Although it's not quite this simple, let's say that there are four broad categories of data storage media drives:

- Magnetic tape on open reels, cartridges, or cassettes
- Optical, including Compact Disk (CD) and Digital Video Disc (DVD)
- Magnetic media in 2.5 and 3.5 inch drive casing - including spinning hard drives
- Solid state including SSD drives in 2.5 and 3.5 inch drive casing, SD cards, and USB flash drives

A few magnetic tape systems may still exist here and there, but the days of laboriously and slowly copying large data sets to banks of multiple backup tapes - and hoping the backup would actually work if you ever needed it - are pretty much over. Trust me: no one is complaining. CDs and DVDs are headed the same direction. Their maximum capacities are no match for the sheer volume of today's enterprise data needs, and consumers don't make local copies of nearly as many large media files as they once did.

Which leaves spinning magnetic and solid state drives.

Gigabyte for gigabyte, spinning hard drives are probably still a bit less expensive than their solid state equivalents (although the price difference is narrowing). But the performance gains delivered by SSDs are very noticeable. Some time ago, I realized that I could actually *save* money by buying smaller capacity SSDs for my personal workstations and laptops instead of larger hard disk drives (HDDs).

Let me explain. The way we use data on our personal computers has changed in recent years. Rather than storing media and software archives locally, we're much more likely to assume they'll be available to stream or download whenever we need them. For most of us, faster download speeds have made "living in the cloud" easy. So we normally just don't need as much storage space any more. The 500GB SSD drive plugged into my busy workstation is barely half full - even taking into account the dozen or so virtual machines and the many ISO images I've got there. And the drive cost me less than I would have paid for a one or two terabyte HDD.

One of the primary roles of storage is data backup. Rather than physically transferring backups between media, local data archiving generally works by moving archives across networks. The trick is designing a backup system that automatically provides you with sufficient duplicates of your archives, rotates them through appropriate life cycles (where, eventually, they're retired and destroyed), and all without generating unnecessary network traffic overhead.

Besides backups, you'll also often want to share data among users working throughout your campus. Two tools for managing both backups and file sharing are network-attached storage (NAS) and storage area networks (SAN). Their similar names suggest they're in the same business. Trust me: they're not.

Network-attached storage (NAS)

NAS is a relatively simple and inexpensive way to share files across a local network. It works through a standalone server device that contains multiple storage drives. The drives will normally be configured as a Redundant Array of Inexpensive Disks (RAID) array to provide redundancy and performance benefits.

The NAS device connects to the network over ethernet cabling and uses regular TCP/IP networking. Client machines in the LAN will see the NAS resources through standard file-sharing protocols like Server Message Block (SMB) and Network File System (NFS).

NAS solutions can be great for smaller home or office environments, but the fun will quickly fade as your infrastructure grows. NAS devices themselves are generally not powerful enough to handle too much of a client workload, and working with large files over an ethernet network may slow things down.

Storage area network (SAN)

If NAS setups are "relatively simple and inexpensive," SANs are complex and expensive. Not by accident were they designed for large enterprise deployments. As a result of the high end hardware you throw into a SAN system, performance will be great and you'll scale much further.

Rather than ethernet, SANs run through much faster Fibre Channel switches (or, sometimes, the slower iSCSI). They provide block based storage rather than file systems and are mounted on client machines as local drives.

Data storage services

As internet connection speeds have improved, it's become more practical to move at least some data archives to the cloud. Instead of local backups - which could be lost in a catastrophic event like a fire - data could regularly be saved to online platforms.

Once there, you'd have a viable, off-site backup. But, if you wanted, you'd also have access to that data from anywhere on earth. If you work remotely with a distributed team, that can be helpful.

You probably already own and have even collaborated on documents that live on Dropbox, Microsoft 365, or Google Drive. For most people, the primary point of interaction for those services is a web browser. But serious data management - or even relatively complex and regular file backups - aren't practical within a browser. So cloud computing providers offer storage and archiving services whose administration can be scripted and automated.

Cloud storage services, like Amazon's Simple Storage Service (S3), provide full archive life cycle management. Data that must remain highly available could, for example, be saved to the S3 Standard storage class.

After a few months - when you're less likely to need the data, but must still retain a copy for regulatory reasons - you could move your archive to the cheaper S3 Glacier class. Data in Glacier is secure and durable, but would take much longer to access if retrieval became necessary. After a full year you might be able to delete it altogether. Better yet, there are simple ways to *automate* the way your data moves through its life cycle.

All major cloud providers will have their own comparable data storage services. Naturally, prices and exact service features will differ from one another. And, of course, feature and pricing details will often change.

It may not always be practical to transfer data to the cloud over the internet. Extremely large data sets can take a very long time to upload even using fast internet connections. Sure, if you're lucky enough to have a fibre optics connection giving you one gigabyte/second, then a one terabyte upload would take only two and a half hours or so (assuming no one else was using the connection).

But what about 100 TBs of data (that'll take you more than ten days)? And what if you only get 100MB/second (more than three months)? Obviously, if you're uploading jumbo-sized archives weekly or have other uses for your internet connection, then uploading isn't an option.

For such cases, you can still get your data into the cloud, but it'll have to find another ride. AWS, as it turns out, offers their Snow Family services. Snowball is a large, secure storage device. It can be safely shipped to AWS customers, loaded up with dozens of terabytes of data, and then shipped back. Once back home at Amazon, the data will be directly uploaded to a bucket in the customer's account. Alternatively, Snowballs can be kept on-location and used as edge compute devices.

Snowball's big brother is AWS Snowmobile, a 45-foot long secure shipping container capable of handling Exabyte-scale digital migration. Snowball's little cousin, AWS Snowcone, is a rugged container the size of a tissue box that can handle eight TB of usable storage, along with the possibility of virtual cloud instances and network connectivity to the AWS cloud. Besides transferring your data, Snowcones can be used as highly mobile edge compute devices in their own right.

10

Working With Data

By now it's no secret that digital data arrives by the truckload and that it can be worth its weight in gold. But that knowledge isn't half as important as understanding *how* you can tame the data beast and then wring out every drop of its value.

Naturally, creative and resourceful people in one place or another are always finding new processes and applications that'll make better use of their data. But we'll explore some of today's dominant data utilization trends and leave predicting tomorrow's technology for the pundits.

Exactly What Is Data?

Before priming our understanding of what's available to help us work productively with data, it's a good idea to first define exactly what data is. Sure, we saw plenty of great individual examples in the previous chapter (Managing Data Storage), including the huge volumes of performance and status information produced by digital components of complex systems like cars. But that's not the same as a definition.

So then let's define it. Data, for our purposes, is any digital information that is generated by, or used for, your compute operations. That will include log messages produced by a compute device, weather information relayed through remote sensors to a server, digital imaging files (like CT, tomography and ultrasound scans), and the numbers you enter into a spreadsheet. And everything in between.

Which brings us to *big data* - another one of those buzz phrases that get thrown around a lot, often without accompanying context or clarity. How would you define big data?

If you answered that *big data* describes data sets that come in volumes higher than traditional data software and hardware solutions are capable of handling, then you're on target. Although we could add one or two secondary characteristics.

The *complexity* of a data set, for instance, is also something that could force you to consider big data solutions. And sets of data that must be consumed and analyzed while in motion (*streaming data*) are also often better addressed using big data tools.

It's worth mentioning that big data workloads will often seek to solve large scale predictive analytics or behavior analytics problems. Such problems are common within domains like healthcare, Internet of Things (IoC), and information technology.

With that out of the way, we can now get to work understanding how - and why - all that data is being used.

Virtual Reality and Augmented Reality

Why? Plain old reality suddenly not good enough for you?

Well yes, in some cases, plain old reality really isn't good enough. At least if you have a strong interest in engaging in experiences that are difficult or impossible under normal conditions.

A virtual reality (VR) device lets you immerse yourself in an non-existent environment. The most common examples of currently available VR technology feature some kind of headset that projects visual images in front of your eyes while tracking your head movements and, in some cases, the way you're moving other parts of your body. The visual images will adapt to your physical movements, giving you the sensation that you're actually within and manipulating the virtual projection.

VR has obvious potential applications in educational, healthcare, research, and military fields. The ability to simulate distant, prohibitively expensive, or theoretical environments can make training more realistic and immediate than would be otherwise possible.

VR technologies have been arriving - and then disappearing - for decades already. For the most part, they've focused on providing immersive gaming and entertainment environments. But they've never really caught on in a big way beyond the niche product level. This might be partly due to high prices, and because some people experienced forms of motion sickness and disorientation.

But maybe - just maybe - <insert the current year here> will finally be the year VR hits the big time.

But where VR can leverage data in a really meaningful way is when, rather than blocking out your physical surroundings, the virtual environment is overlaid *on top* of your actual field of vision. Imagine you're a technician working on electrical switching hardware under a sidewalk. You're wearing goggles that let you see the equipment in front of you, but that also project text and icons clearly identifying labels for each part and that show you where a replacement part should go and how it's connected. This is *augmented reality.*

I'm sure you can easily imagine how powerful this kind of dynamic display could be in the right conditions. Surgeons are able to access a patient's history or even consult relevant medical literature without having to divert their eyes from the operation. Military pilots can similarly enjoy "heads up" displays that show them timely status reports describing their own aircraft and broader air traffic conditions without distraction.

Artificial Intelligence and Machine Learning

As a rule, computers are even better at performing dull, repetitive tasks over and over again than bored teenagers pretending to do homework. And they don't need parents to clean up after them.

The trick with computers is to cleverly string lots of dull, repetitive tasks together so that they can approximate intelligent and useful behavior. The prize at the end of that road is called *automation.* Or, in other words, a state where computers can be confidently left alone to perform complex and useful tasks without supervision.

In many ways, we've been living in an age of sophisticated computer automation for decades. Domains as diverse as security monitoring, urban traffic control, book manufacturing, and heavy industry are already being handled with little or no human supervision. But artificial intelligence (AI) seeks to go beyond relatively simple repetition to train computers to think for themselves - and thereby efficiently solve far more difficult problems.

Great idea. Somewhat harder to achieve in the real world.

What can AI actually do?

Understanding how effective AI can be will depend on what you expect it to do. Can you design software to search for and flag a handful of suspicious financial transactions from among the millions of credit card transactions a large bank processes? Yes. Although I'm not quite sure that's truly *AI* at work and not just automation.

Can you deploy "intelligent" chatbots on your website to help customers solve their problems without needing actual (and expensive) human interaction? Yes. In fact, I just had a surprisingly effective conversation with my mobile phone carrier's chatbot that did quickly solve my problem.

Can the first stages of a rocket you've just used to launch a payload into space use AI to guide it to a safe landing on a moving platform in the middle of the ocean? If you'd ask me, I'd say it's impossible. But SpaceX went ahead anyway and did it multiple times. Good thing they didn't ask me.

But can AI reliably make strategic decisions that intelligently account for all the many moving parts and complexity that exist in your industry? Can an AI-powered machine pass the Turing test (where a human evaluator is unable to be sure whether the machine is also human)? Perhaps not just yet. And perhaps never.

One tool used in many AI processes is the neural network. The original neural network consists of the many neurons that carry information about the state of a biological environment to the brain. Artificial and virtual neural networks are systems for assessing, processing, and responding to the large physical or virtual data sets that feed AI-controlled systems. Such data can come from cameras or other physical sensors, or from multiple data sources. The processed data can sometimes be used for predictive modeling, where the likelihood of future outcomes are compared.

Exciting stuff, to be sure. But the tools used for some of the most significant accomplishments attributed to artificial intelligence aren't actually artificial. Nor did they necessarily require all that much intelligence.

For example, Amazon Mechanical Turk (MTurk) is a service that connects client companies with remote freelancing "human intelligence" workers. The workers will, for what usually amounts to dreadfully low pay, perform "mechanical" tasks like labeling the content of hundreds or thousands of images. The labeling will cover areas like "is the subject a male or female?" or "is the subject a car or a bus?"

It could be that, over time, services like Mechanical Turk will become less important as improving AI methodologies might one day completely replace the human element for this kind of work. But in the meantime, MTurk and its competitors are still steaming along at full speed, churning out millions of units of "artificial" artificial intelligence.

One methodology that can help reduce reliance on human intervention is machine learning (ML).

How can machine learning help?

ML works by leveraging various kinds of manual assistance to help achieve greater task automation. An ML system can hopefully "learn" how to manage our tasks by being exposed to existing training data. Only once the system has demonstrated sufficient skill at solving the problems you have for it, will it be let loose on "real world" data.

These are some common approaches to training your ML system:

- **Supervised learning** lets the ML software read data sets that include both "problems" (images, for example) and their "solutions" (full labels). By seeing enough of the provided examples, the system should be able to apply its experience to similar problems that arrive without solutions.

- **Unsupervised learning** simply throws raw data without any associated solutions at the system. The goal is for the software to recognize enough patterns in the data to allow it to solve the problems on its own.
- **Reinforcement learning** learns from interactions with its environment. Ideally, the software recognizes and understands positive results and evolves its methodology to reliably and consistently produce similar results.
- **Deep learning** algorithms apply multiple layers of analysis to transform the raw target data. The full, multi-layer process in deep learning is known as the *substantial credit assignment path* (CAP).

AI in general, and ML in particular, are effective at building tools for tasks like autonomous driving, drug discovery, email filtering, and speech recognition, and for deriving sentiment analysis from massive data sets made up of human communications.